United States
Environmental Protection
Agency

Enforcement and
Compliance Assurance
(2221-A)

EPA 305-B-99-005
March 1999

I0502985

Multimedia Environmental Compliance Guide for Food Processors

SECTION 1 CONTENTS

1. THE GUIDE: WHAT IT IS; WHAT IT DOES

1.1 Why an Environmental Compliance Guide for Food Processors

As food processors, you are regulated by a variety of federal laws administered by the U.S. Environmental Protection Agency (EPA) that impact human activities and the environment. Noncompliance with these regulations can damage human health and the environment, and result in significant financial liabilities for clean up costs or fines. Environmental compliance may be difficult for some food processors that do not have the time, staff, or other resources necessary to determine their responsibilities. Also, environmental regulations and laws can be complicated, and information on environmental compliance may be difficult to locate. Adding to these complexities, you must be aware of and meet stringent food safety requirements. To assist you, EPA, with special assistance from the American Frozen Food Institute (AFFI), the American Meat Institute (AMI), the National Food Processors Association (NFPA) and the Food Industry Environmental Council (FIEC), has developed this guide to address these issues.

1.2 How to Use This Guide

This guide is intended to provide you, the owner/operator, with a good **first step** in understanding EPA's environmental requirements affecting your specific operations. The guide explains the basis of EPA's major statutes and provides a general overview of a food processor's major EPA requirements. The requirements discussed here should provide a good framework for understanding your federal environmental compliance responsibilities, **but this guide does not provide the final word on what your compliance responsibilities are and how you meet them.** You should consult directly EPA's regulations, program guidance, and other compliance assistance materials.

> ***State/Local Requirements:*** *The regulations discussed in this guide are primarily EPA requirements. Your state may have its own, stricter requirements; however, state regulations usually are based on federal law. Be sure to check your state and/or local government environmental requirements.[1]*
>
> ---
>
> [1] ***The Source Book of State Laws & Regulations for Food Processors****, published by the NFPA in June 1996, offers help in understanding the regulatory requirements in each of the 50 states.*

Organization

The guide begins with a brief overview of the industry and an introduction to several important EPA policies and systems that are designed to foster environmental compliance. Following this introduction, a brief summary of the major environmental statutes applicable to the food processing industry is presented in Section 2.0 *Guide to EPA's Major Environmental Statutes*. Next, to assist you in identifying EPA's statutes and regulations applicable at your facility, Section 3.0 *Understanding the Process: Inputs, Outputs, and Applicable Federal Environmental*

Regulations presents a method of identifying (1) your facility's wastestreams (regulated outputs) and (2) hazardous or other regulated materials that may be inputs to either your process or to ancillary operations (e.g., refrigeration).

Section 3.0 also helps you identify which EPA statute(s) may apply to your inputs and wastes. With this information, you can then refer to the following sections of the guide to learn more about applicable EPA regulations and requirements:

- Section 4.0   How Do I Comply With Wastewater Discharge and Related Regulations?
- Section 5.0   How Do I Comply With Safe Drinking Water Regulations?
- Section 6.0   How Do I Comply With Air Regulations?
- Section 7.0   How Do I Comply With the Emergency Planning and Community Right-to-Know Act Regulations?
- Section 8.0   How Do I Comply With the Hazardous Waste Regulations?
- Section 9.0   How Do I Comply with Spill or Chemical Release Requirements?
- Section 10.0   Other Major Environmental Statutes and Regulations: CERCLA, RCRA Subtitle D, FIFRA, and TSCA.

Remember that this guide highlights major EPA requirements only. To help you in understanding the full range of EPA requirements, Appendix A. *Summary of Major Regulations From the CFR* provides relevant portions of the Code of Federal Regulations (CFR) in an easier-to-understand format (with CFR citations). If you have additional questions, Appendix B. *Resources* provides state and federal agency contacts, hotlines, and an annotated list of important EPA policies and guidance. References associated with this guide are presented in Appendix C. *References*.

> *Your state may have its own, stricter requirements than the federal requirements. Be sure to contact your state regulatory agency for information on state requirements.*

Attention to Food Safety

As you are well aware, food safety is a paramount objective of food processors and agencies which regulate the industry. Food safety should be kept in mind when reviewing the information in this guide and complying with environmental regulations. Under the Federal Food, Drug and Cosmetic Act (FFDCA), the Food and Drug Administration (FDA) of the U.S. Department of Health and Human Services (USDHHS), and the Food Safety and Inspection Service (FSIS) of the U.S. Department of Agriculture (USDA) regulate the food processing industry to assure the safety of the food supply. These regulations address sanitation, microbial pathogens, and other sources of foodborne illness. EPA also is involved in food safety by virtue of its responsibility under the Federal Insecticide, Fungicide, and Rodenticide Act (FIFRA) to set pesticide tolerances or maximum pesticide residue levels. Some recent activity related to food safety which may directly affect food processors includes the following:

- In 1996, Congress amended both FFDCA and FIFRA under the Food Quality Protection Act (FQPA) to restructure the standard setting, review, and enforcement authorities for pesticide tolerance and residues in food. See Section 10.3 *Federal Insecticide, Fungicide, and Rodenticide Act* for more information.

- On May 12, 1997, Vice President Al Gore announced a five-point Administration plan to strengthen and improve food safety, known as the *National Food Safety Initiative*

Program. The plan sets forth new steps to reduce foodborne illness from microbial contaminants by working with consumers, producers, industry, states, universities, and the public.

Ideas for Pollution Prevention

Many industry sectors are experimenting with new approaches to environmental compliance through pollution prevention (P2) techniques, such as reducing the use of hazardous chemicals by switching to alternative less hazardous substances, recycling materials, and reducing wastewater discharge by means of water conservation. The final section of this guide, Section 11.0 *Pollution Prevention Techniques*, contains general information about means of reducing or eliminating your wastestreams. Cost effective compliance and pollution prevention techniques are discussed further in Section 1.5.

1.3   Tools That Encourage Environmental Compliance

1.3.1 Compliance Incentives and Policies

Since 1986, EPA has promoted the use of environmental auditing by companies in the United States. In 1995, EPA updated and expanded its earlier policy in the form of two new policies that encourage companies to achieve environmental compliance by providing incentives for them to conduct environmental audits voluntarily (see *Incentives for Self-Policing Policy*) and to participate in onsite compliance assistance programs *(see Small Business Policy)*. These policies focus on proactive identification and prompt correction of violations, and provide penalty mitigation for those who qualify.

> *Environmental Audit: For purposes of these policies, an environmental audit is defined as "a systematic, documented, periodic, and objective review by regulated entities of facility operations and practices related to meeting environmental requirements."*

- *Incentives for Self-Policing Policy*. In 1995, EPA issued *Incentives for Self-Policing: Discovery, Disclosure, Correction, and Prevention of Violations* (commonly known as the *Audit Policy*). Under this policy, companies that (1) voluntarily discover, promptly disclose, and correct violations; (2) prevent their recurrence; and (3) promptly remedy any resulting damage do **not** face gravity-based penalties (i.e., the penalty amount over and above the company's economic gain from noncompliance). EPA retains its discretion to recover **economic benefit** gained as a result of noncompliance, so that companies will not be able to obtain an economic advantage over their competitors by delaying their investment in compliance. EPA continues its practice of not routinely requesting environmental audit reports. For additional information on this policy and how it may apply to you, you can review the following: the final policy in the *Federal Register* (60 FR 66706-66711 [December 22, 1995]); the *Audit Policy Interpretive Guidance*; and the *Audit Policy Update*, EPA's newsletter about the implementation of the policy. All three are available at the website of EPA's Office of Enforcement and Compliance Assurance (OECA) at **http://es.epa.gov/oeca/auditpol.html/**.

- **Small Business Policy**. EPA's *Policy on Compliance Incentives for Small Businesses* (*Small Business Policy*), issued on May 20, 1996, was developed to help small businesses achieve environmental compliance. If you are a small business with 100 or fewer employees in the company, you may be eligible to have all potential penalties for noncompliance waived, if you agree to come into compliance and meet other criteria.

> **Definition:** *Any business owned by a person, corporation, or partnership that employs 100 or fewer individuals across all facilities and operations owned by the entity is considered a small business.*

The policy provides incentives such as **penalty waivers** or **penalty reduction** for businesses that participate in government-sponsored, onsite compliance assistance programs, or that conduct environmental audits to discover, disclose, and correct violations. Onsite compliance assistance includes information or assistance provided by EPA, a state agency, or other government agency or government-supported entity during a visit to your facility to help you comply with mandated environmental requirements. This assistance may be obtained confidentially from those state assistance programs that offer such confidentiality.

The sooner **noncompliance** is disclosed and corrected, the greater the benefits. A penalty can be waived entirely or reduced significantly if violations are self-disclosed to the proper authorities and corrected. If compliance is delayed, however, and EPA discovers the violation on its own, the ensuing penalties may be very costly. You can review the final policy [*Federal Register* 61 FR 27984; June 3, 1996], and more information on how it may apply to your business by accessing **http://es.inel.gov/oeca/smbusi.html/**.

To assist you with compliance, EPA and states have developed numerous guides written in "plain English" on how to conduct audits and understand regulations. Contact the Small Business Ombudsman in your EPA regional office or visit EPA's Internet site for more information at **http://www.epa.gov/ttnsbap1/**--*Small Business Assistance Programs*. You also may call the EPA's National Small Business Ombudsman's toll-free hotline at 1-800-368-5888.

Many states have adopted incentives for environmental auditing -- some through policies and others by legislation. Some states have adopted laws that provide broad privileges and immunities for environmental audit findings. EPA opposes these laws and is working with state officials to resolve issues with regard to the state maintaining necessary enforcement and information gathering authorities, and ensuring legally mandated public access to information.

To the extent that violations, revealed through an audit and disclosed under a state audit privilege and immunity law, continue to be addressed inadequately, EPA may take action through its oversight authorities granted under the federal environmental laws. However, EPA has not brought, and will not bring enforcement actions against companies merely because they take advantage of state audit privilege and immunity laws.

1.3.2 Environmental Management Systems

An **environmental management system (EMS)** uses a defined process to identify the environmental impacts of your operations, set goals, implement procedures to minimize those impacts, and measure results to determine whether established goals and procedures are appropriate. Such a system has the potential to improve your company's environmental performance and compliance with regulatory requirements. It also may save your company money, reduce liability, and improve efficiency in operations. An example in the food processing industry is the environmental operating plan that Jack M Berry, Inc. (LaBelle, FL), a mid-sized juice-processing facility, is developing under EPA's Project XL (X for environmental eXcellence and L for Leadership). See Section 11.4.1 *EPA Programs* for a brief explanation of this effort.

> *At this time, EPA is not basing any regulatory incentives solely on the use of EMSs, or certification to ISO 14001. For more information on EPA's position, see the Federal Register Notice, EPA Position Statement on Environmental Management Systems and ISO 14001 and a Request for Comments on the Nature of the Data to be Collected from Environmental Management System/ISO 14001 Pilots, 63 FR 12094-97, March 12, 1998, from the EPA's Office of Reinvention. This notice can be accessed at EPA's website at **http://www.epa.gov/reinvent/ notebook/emsfr1.htm/.***

EPA supports the development and use of EMSs that help a business achieve its environmental obligations and broader environmental performance goals. EPA encourages the use of EMSs that focus on improved environmental performance and compliance, as well as source reduction (pollution prevention) and system performance. By working in partnership with a number of states, EPA is exploring the utility of EMSs, especially those based substantially on ISO 14001. ISO 14001, an international standard finalized in 1996 by the International Organization for Standardization (ISO), is based on previous standards and agreement by international business and government representatives.

An EMS includes five key elements which are defined as follows:

(1) The **environmental policy** sets the general direction for your company's EMS, and if appropriately communicated, shows management's commitment that facilitates implementation of the EMS throughout all levels of your organization.

(2) **Planning** means examining the environmental aspects of your operations more closely. Based on this review, you can develop objectives and targets designed to minimize environmental impacts and improve the overall performance of your company. When you complete the planning process, your organization will have defined the objectives of its environmental program, and developed a plan to meet them.

> *When planning for an EMS, you might examine water usage. Based on this review, you may decide to reduce water usage by installing low flow nozzles.*

Through the remaining components, (3) **implementation and operation**, (4) **checking and corrective action**, and (5) **management review**, your company develops mechanisms to achieve these objectives, reduce catastrophic risk, and continually monitor its environmental activities. These mechanisms are specific to your company's operations, and are reviewed and revised to promote continuous improvement.

EPA recognizes the potential value of a mature EMS. Through initiatives such as Project XL, EPA is working with companies to test EMSs that are designed to achieve superior environmental performance. In 1994, through its Environmental Leadership Program (ELP), EPA issued guidance on EMSs, entitled *Draft Program Guide: Appendix A - ELP Environmental Management System Guidelines.* Although developed for use by facilities applying to ELP, the criteria in the document may be useful to your company. You can access this document through the ELP Homepage at **http://es.epa.gov/elp/append-a.html/**. See Section 11.4.1 *EPA Programs* in this guide for a brief explanation of both Project XL and ELP.

1.4 Brief Overview of the Food Processing Industry

The food processing products industry is a manufacturing industry that processes raw or prepared animal, marine, and vegetable materials into intermediate foods or edible products. This industry includes establishments manufacturing or processing foods and beverages for human consumption, and certain related products, such as manufactured ice, chewing gum, vegetable and animal fats and oils, and prepared feeds for animals and fowls. Processes of this industry result in the conversion of bulky, perishable, or inedible food materials into more palatable, or more convenient foods and beverages.[1]

Table 1-1 lists the Standard Industrial Classification (SIC) and the North American Industrial Classification System (NAICS) codes for the types of establishments (i.e., facilities) as defined by the U.S. Census Bureau within SIC Code 20.

As of the Fall 1998, EPA has not published an overall plan for phasing in NAICS codes. But several EPA programs that rely on SIC codes in determining whether a regulatory requirement applies to a given facility have begun to adopt the NAICS codes. The EPCRA Non-313 Program and the Oil Pollution Prevention Program proposed such changes in regulatory criteria in the Federal Register in 1998. Also, the EPCRA 313 Toxic Release Inventory Program has begun planning how to incorporate NAICS codes into its regulatory criteria. This may be a several year effort. For additional information about these respective changes, contact EPA's Chemical Emergency Preparedness and Prevention Office (CEPPO); the Oil Program Center (OPC); or the Office of Prevention, Pesticides and Toxics (OPPT). Information about the EPCRA requirements and Internet sites can be found in Section 7.0 *How Do I Comply With the Emergency Planning and Community Right-to-Know Act Regulations?*; information about the Oil Pollution Prevention Program requirements and Internet site can be found in Section 4.6 *Oil Pollution Prevention Regulation* of this guide.

---

[1] U.S. Environmental Protection Agency. *Industry Profiles: Food and Kindred Products and Stone, Clay, Glass, and Concrete.* Office of Solid Waste. Prepared by ICF, Inc., July 1994.

Table 1-1. SIC and NAICS Codes for the Food Processing Industry*

SIC Codes	Types of Facilities	NAICS Codes	
201	Meat products	311611 311612 311615, 311999	Meat packaging plants Sausages and other prepared meat products Poultry slaughtering and processing
202	Dairy products	311512 311513 311514 31152 311511	Creamery butter Natural, processed, and imitation cheese Dry, condensed, and evaporated dairy products Ice cream and frozen desserts Fluid milk
203	Canned, frozen, and preserved fruits, vegetables, and food specialties	311422, 311999 311421 311423 311421, 311941 311411 311412	Canned specialties Canned fruits, vegetables, preserves, jams, and jellies Dried and dehydrated fruits, vegetables, and soup mixes Pickled fruits and vegetables, vegetable sauces and seasonings, and salad dressings Frozen fruits, fruit juices, and vegetables Frozen specialties not elsewhere classified
204	Grain mill products	311211 31192, 31193 311212 311822 311221 311111 311611, 311119	Flour and other grain mill products Cereal breakfast foods Rice milling Prepared flour mixes and doughs Wet corn milling Dog and cat food Prepared feeds and feed ingredients for animals and fowls, except dogs and cats
205	Bakery products	311812 311821, 311919, 311812 311813	Bread and other bakery products, except cookies and crackers Cookies and crackers Frozen bakery products, except bread
206	Sugar and confectionary products	311311 311312 311313 31133, 31134 31132 31134 311911	Cane sugar, except refining Cane sugar refining Beet sugar Candy and other confectionary products Chocolate and cocoa products Chewing gum Salted and roasted nuts and seeds

Table 1-1. SIC and NAICS Codes for the Food Processing Industry*

SIC		NAICS	
SIC Codes	Types of Facilities	NAICS Codes	Types of Facilities
207	Fats and oils	311223, 311225 311222, 311225 311223. 311225 311613, 311711, 311712, 311225 311225, 311222, 311223	Cottonseed oil mills Soybean oil mills Vegetable oil mills, except corn, cottonseed, and soybeans Animal and marine fats and oils Shortening, table oils, margarine, and other edible fats and oils, not elsewhere classified
208	Beverages	31212 311213 31213 31214 312111, 312112 31193, 311942, 311999	Malt beverages Malt Wines, brandy, and brandy spirits Distilled and blended liquors Bottled and canned soft drinks and carbonated waters Flavoring extracts and flavoring syrups, not elsewhere classified
209	Miscellaneous food preparations and kindred products	311711 311712 31192, 311942 311919 312113 311823 311423, 111998, 31134, 311991, 31183, 31192, 311941, 311942, 311999	Canned and cured fish and seafoods Prepared fresh or frozen fish and seafoods Roasted coffee Potato chips, corn chips, and similar snacks Manufactured ice Macaroni, spaghetti, vermicelli, and noodles Food preparations, not elsewhere classified

*   Table 1-1 lists the 1987 SIC codes and 1997 NAICS codes for the food processing industry. The 1997 NAICS codes will replace the 1987 SIC codes in publications of the U.S. Statistical Agencies over several years (1998-2004), beginning with publications of the NAICS United States Manual. The NAICS Implementation Schedule for these agencies is available on the U.S. Census Bureau's Internet site at **http://www.census.gov/epcd/naics/ timeschd.html/.**

Encompassing all facilities in SIC group 20, titled Food and Kindred Products, this is one of the largest industry groups comprising the manufacturing sector of the U.S. economy.[2] Figure 1-1, shows the distribution of food processing facilities across the U.S. in 1994. The four states with the largest number of establishments were:

- California: 2,721 establishments (13.1%)
- New York: 1,363 establishments (6.5%)
- Texas: 1,187 establishments (5.7%)
- Pennsylvania: 1,111 establishments (5.3%).

---

[2] The term "food processing" is used throughout the guide in place of Food and Kindred Products.

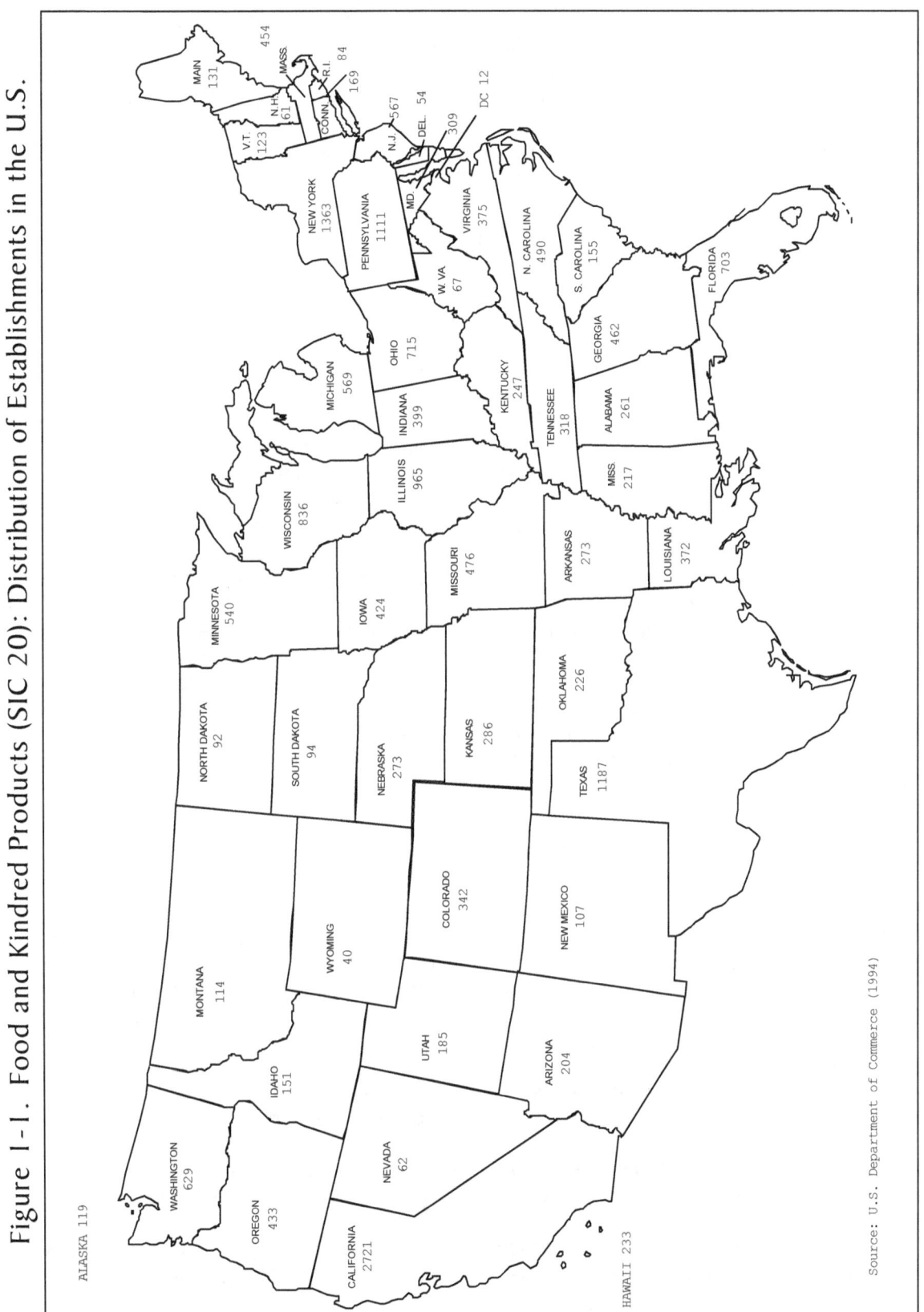

Figure 1-1. Food and Kindred Products (SIC 20): Distribution of Establishments in the U.S.

Source: U.S. Department of Commerce (1994)

Together, these states contain approximately 30 percent of the establishments nationwide. According to the 1994 Census of Manufacturers, there were 20,800 establishments in SIC 20 with shipments valued at $430 billion. In terms of employment, the food processing industry ranks fourth in the Nation in 1994, providing 1.5 million jobs. According to the data in EPA's Toxic Release Inventory (TRI) system, the chemicals released from the food processing industry that may have environmental impact include ammonia, phosphoric acid, sulfuric acid, chlorine, hydrochloric acid, nitric acid, copper compounds, and zinc compounds.

1.5   Cost Effective Compliance and Pollution Prevention Techniques

Understanding federal, state, and local environmental requirements is the first step to cost effective compliance. Finding the most effective means of environmental compliance and going beyond compliance are the next steps. This involves knowing what your compliance activities are and how much you are spending on them (in relative terms at least). This involves establishing a baseline of how much your facility is spending to comply with environmental regulations. With this type of information, you can begin to assess where you might save money through more effective means of complying or through pollution prevention.

Compliance activities for your facility involve interactions between you and the regulatory agencies. These activities include regulatory obligations such as obtaining permits, paying fees, monitoring, and reporting. Planning activities, such as emergency response, and recordkeeping also may be required which may have additional costs. And, of course, pollution control may involve both capital and operating expenses. Depending on the wastestreams your facility generates and how you manage those wastes, you may be subject to this array of requirements. Pollution prevention techniques may help you cut costs, and, in some instances, may enable your operation to drop below regulatory thresholds and thereby become free of certain regulatory requirements.

What is Pollution Prevention?

Pollution prevention (P2) encompasses both source reduction and in-process recycling. The federal Pollution Prevention Act of 1990 defines source reduction as any practice that reduces the amount of any hazardous substance, pollutant, or contaminant entering any wastestream (including fugitive emissions) prior to recycling, treatment, or disposal, and that reduces the hazards to public health and the environment associated with the release of such substances, pollutants, or contaminants. The Act declares that governments, businesses and industries, and individuals should prevent or reduce pollution at its source wherever feasible. Where source reduction cannot be achieved, the Act advocates that responsible parties reuse and recycle to reduce the quantity of hazardous waste requiring treatment.

EPA has adopted a hierarchical approach to environmental protection, including source reduction and in-process recycling, as follows:

   A. ***Source Reduction.*** The most desirable option of the hierarchy and the most effective way to reduce risk is through source reduction. Source reduction is defined as any method that reduces or eliminates the source of pollution entirely. This includes any practice that:

- Reduces the amount of hazardous substances, pollutants, or contaminants that enter a wastestream or are released otherwise into the environment prior to recycling, treatment, or disposal.

- Reduces the hazards to public health and the environment associated with the release of such substances, pollutants, or contaminants.

B. *Recycling.* Where pollution cannot be prevented through source reduction methods, the wastes contributing to the pollution should be recycled, preferably through closed loop, in-process or in-line methods. Other recycling approaches include the use, reuse, or reclamation of waste after it has been generated.

C. *Treatment.* Wastes that cannot be feasibly reduced at the source or recycled should be minimized through treatment in accordance with environmental standards that are designed to reduce both the hazard and volume of wastestreams.

D. *Disposal.* Finally, any residues remaining from facility operations that cannot be reduced, reused, recycled or treated should be disposed of safely to minimize their potential for release into the environment. Disposal involves the transfer of a pollutant to the environment in either air, solid waste, or water.

Section 11.0 *Pollution Prevention Techniques* provides general information and references on P2 techniques you might implement at your facility. Some are easy, some are more challenging, and they all involve changes in how you do business. EPA encourages pollution prevention as a solution.

Keep in mind that there may be state pollution prevention requirements with which you must comply. Contact your state regulatory agency for more information. Also, carry out all pollution prevention activities at your facility in accordance with food safety requirements of the USDA and FDA.

SECTION 2 CONTENTS

2. GUIDE TO EPA'S MAJOR ENVIRONMENTAL STATUTES

2.1 Introduction

This section discusses the federal statutes administered by U.S. Environmental Protection Agency (EPA) along with the citations to EPA regulations that may apply to the food processing sector. For a brief discussion of pending and proposed EPA regulations that may apply to the food processors, consult Appendix A.7 *Pending and Proposed Regulations* of this guide.

The descriptions within Section 2.0 are intended solely for general information. Depending upon the nature or scope of the activities at a particular facility, these summaries may or may not necessarily describe all applicable environmental requirements. Moreover, they do not constitute formal interpretations or clarifications of the statutes and regulations. You must keep in mind that this section discusses federal environmental laws administered by EPA (with a few exceptions) and their related regulations only. State and/or local regulations may be more stringent than federal regulations. Therefore, it is essential to consult appropriate state and/or local agencies to learn the full range of environmental requirements that may aply to your facility.

For further information about federal requirements, you should consult the Code of Federal Regulation (CFR) citation(s) listed at the end of each section, and review Appendix A. *Summary of Major Regulations From the CFR* that contains summaries of the portions of the CFR pertaining to each major statute discussed in the guide. Appendix B. *Resources* lists regional and state regulatory agency contacts and EPA hotlines. Appendix C. *References* lists documents used in developing this guide and additional references for your use.

2.2 Clean Water Act (CWA) and Oil Pollution Act (OPA)

The discharge of wastewater from your food processing facility generally will be covered by either the federal Clean Water Act (CWA) or the Safe Drinking Water Act (SDWA) (see Section 5.5 *Underground Injection Control*). In 1972, Congress passed the Federal Water Pollution Control Act (FWPCA), now known as the CWA, which established the basic framework for protecting the waters of the United States. The CWA and its regulations now focus on keeping conventional, nonconventional (including oil and grease), and toxic water pollutants out of our rivers, lakes, and oceans.

> *Wastewater discharge requirements are discussed in more detail in Section 4.0 and Appendix A.1. For information on planning and release reporting requirements for spills to water, see Section 9.0.*

Generally, federal regulations target three types of industrial discharges. Industrial wastewater discharges from food processing facilities probably fall into one of these categories:

(1)   **Direct discharges** which include any wastewater from an industrial facility (e.g., untreated, unpolluted wastewater or treated process wastewater) that is discharged straight to surface waters (e.g., ponds, lakes, oceans, streams, and wetlands). Storm water discharges also are considered a type of direct discharge.

(2)   **Indirect discharges** which include any wastewater from an industrial facility that is discharged to a publicly owned treatment works (POTW), which subsequently discharges to a surface water.

(3)   **Land application** of industrial wastewater discharges. Land application discharges include any wastewater from an industrial facility that is discharged to land to either condition the soil or to fertilize crops or other vegetation grown in the soil.

For more information, see:

- Section 4.0 *How Do I Comply With Wastewater Discharge and Related Regulations?*
- 40 CFR 110-122, 40 CFR 400-500: Clean Water Act and Effluent Guidelines
- •• Appendix A.1. *Summary of Principal Regulations Under the Clean Water Act.*

NPDES Permit Requirements. The National Pollutant Discharge Elimination System (NPDES) program (CWA Section 402) controls direct discharges into navigable waters.   A NPDES permit sets limits, often referred to as **effluent limits** on the amounts of pollutants that can be discharged to surface waters.

> *NPDES requirements are discussed in more detail in Section 4.3 and Appendix A.1.*

Permits must be obtained from EPA or the authorized state or territory. As of March 1998, EPA has authorized 42 states and one territory to administer the NPDES program. EPA has not delegated authority to the following states and territories: Alaska, Arizona, District of Columbia, Idaho, Maine, Massachusetts, New Hampshire, New Mexico, Pacific Territories, Puerto Rico, Texas, and the federal Tribal Lands.

For more information, see:

- Section 4.3 *Am I A Direct Discharger?*
- 40 CFR 122: National Pollutant Discharge Elimination System Permit Requirements
- Appendix A.1. *Summary of Principal Regulations Under the Clean Water Act.*

Storm Water Discharges. In 1987, the CWA was amended to require EPA to establish a program to address **storm water discharges** as point sources. Under Phase I of the storm water program, which currently is being implemented, storm water discharges

> *Storm water requirements are discussed in more detail in Section 4.3.2 and Appendix A.1.*

associated with industrial activity, such as food processing, must be covered by a NPDES storm water permit regardless of whether they discharge to a municipal separate storm sewer system or directly to waters of the United States. These permits provide a mechanism for monitoring the discharge of pollutants from these sources to waters of the United States and for establishing appropriate controls. The term "storm water discharge associated with industrial activity" means a storm water discharge from one of 11 categories of industrial activity defined in 40 CFR 122.26. See Section 4.3.2 *Storm Water Dischargers* for more information on these categories and how they apply to your facility.

Food processors that have **no exposure of materials and activities to storm water** are exempt from these requirements. **No exposure** means that there is **no possibility** of storm water, snow fall, snow melt, or storm water "run on" coming in contact with any process or storage related activity. Additionally, storm water permits are not required where runoff flows through a combined sewer to a POTW.

Facilities can comply with NPDES permit requirements for storm water discharges by submitting (1) a Notice of Intent (NOI) to be covered under a **general** permit (Baseline or Multi-Sector); or (2) an application for an **individual** permit; and (3) complying with all of the conditions specified in the applicable permit. In the past, facilities could submit an application to be covered under a group permit, but this option and the original group permit have expired. As of March 1998, 42 states and one territory have been delegated authority by EPA to administer the NPDES program. EPA has not delegated authority to the following states and territories: Alaska, Arizona, District of Columbia, Idaho, Maine, Massachusetts, New Hampshire, New Mexico, Pacific Territories, Puerto Rico, Texas, and the federal Tribal Lands. Of the delegated NPDES states and territories, only the Virgin Islands has not been delegated authority for the storm water general permits program as well.

As part of the storm water permits, facilities are required to develop and implement storm water pollution prevention plans (SWPPPs). These plans are intended to prevent storm water from coming in contact with potential contaminants. Each plan is facility specific because every facility is unique in its source, type and volume of contaminated storm water discharges. Regardless of the variations, all plans must include several common elements, such as a map and site specific considerations.

For more information, see:

- Section 4.3.2 *Storm Water Dischargers*
- 40 CFR 122.26: Storm Water Discharges
- • Appendix A.1. *Summary of Principal Regulations Under the Clean Water Act.*

Pretreatment Program. Industrial wastewater that is treated by a POTW is another type of discharge that is regulated by the CWA. The national **pretreatment program** (CWA 307(b)) controls the indirect discharge of pollutants to POTWs by

Pretreatment requirements are discussed in more detail in Section 4.4.1 and Appendix A.1.

"industrial users." Facilities regulated under 307(b) must meet certain pretreatment standards. The goal of the pretreatment program is to protect the following: (1) municipal wastewater treatment plants from damage that may occur when hazardous, toxic, or other wastes are discharged into a sewer system from industrial activities; (2) the quality of sludge generated by these plants; and (3) the receiving water by preventing the introduction of pollutants into POTWs which will pass through the treatment works. Discharges to a POTW are regulated primarily by the POTW itself, rather than the state or EPA. New food processing facilities, or facilities that have not contacted their POTW in the past, can find the number of their local POTW in the phone book. The wastewater treatment superintendent or pretreatment coordinator can provide the facility with any necessary information.

There are three types of pretreatment requirements: requirements for general industry (**general pretreatment standards**), requirements for specific industries (**categorical pretreatment standards**), and locally established requirements for specific facilities (**local limits**). EPA does not consider food processing facilities to be categorical users and has not established specific numerical limits for indirect discharges from food processors. Hence, **categorical pretreatment standards** that apply to food processing operations require compliance with 40 CFR 403 **(general pretreatment standards)**. **Local limits** may also apply to your facility.

For more information, see:

- Section 4.4 *Am I An Indirect Discharger?*
- 40 CFR 403: Pretreatment Program Requirements
- Appendix A.1. *Summary of Principal Regulations Under the Clean Water Act.*

Oil Pollution Prevention Requirements. In 1973, EPA issued the Oil Pollution Prevention Regulation (40 CFR 112), also known as the Spill Prevention, Control and Countermeasures (SPCC) regulation, to address the oil spill prevention provisions contained in the Clean Water Act (CWA) of 1972. The main objective of the SPCC program is

> *Oil Pollution Prevention requirements are discussed in more detail in Section 4.6 and Appendix A.1.*

to **prevent** oil spills from **regulated aboveground and underground storage tanks** from reaching navigable waters of the U.S. or adjoining shorelines. In 1990, Congress passed the Oil Pollution Act (OPA) that amended Section 311 of the CWA to require **substantial harm** facilities to develop and implement facility response plans (FRPs). FRPs help facility owners/operators develop a response organization and identify the resources needed to respond to an oil spill adequately and in a timely manner.

Under the CWA, the definition of oil includes oil of any kind and any form, such as petroleum and nonpetroleum oils. Generally, oils fall into the following categories: crude oil and refined petroleum products, edible animal and vegetable oil, other oils of animal or vegetable origin, and other nonpetroleum oils.

EPA's regulation requires facilities to prepare a plan and implement measures to prevent and control oil spills, regardless of the cause (e.g., human operational error, equipment failure or natural causes, such as lightning striking a tank). If your facility is subject to the SPCC requirements, EPA requires you to prepare an SPCC plan and conduct an initial screening to determine whether you are required to develop an FRP. Those facilities that could cause **substantial harm** to the environment must prepare and submit an FRP to EPA for review.

In the event of an oil spill or release, you must first report it to the **National Response Center at 1-800-424-8802 or 703-412-9810 (Washington, D.C. area)**. In addition, you (the owner or operator of a regulated facility) must submit, in writing, certain information including the SPCC Plan to the EPA Regional Administrator within 60 days, if the release meets either of the following conditions: (1) **either** a single discharge of more than 1,000 gallons of oil; **or** (2) two reportable spills/discharges of oil in harmful quantities, during any 12-month period, into or upon navigable waters, shorelines, etc.

For more information, see:

- Section 4.0 *How Do I Comply With Wastewater Discharge and Related Regulations?*
- 40 CFR 112 Oil Pollution Prevention Regulation
- Appendix A.1. *Summary of Principal Regulations Under the Clean Water Act.*

Examples of CWA Enforcement Provisions and Penalties

The 1987 amendments to the CWA increased EPA's penalty authorities as an enforcement tool. Congress added new authority for assessment of administrative penalties and increased penalties for civil and criminal violations. Some examples of EPA's enforcement authorities under the NPDES Program and the Oil Pollution Prevention Program are summarized below. Civil penalty amounts presented here also reflect the inflation adjustment authorized by Congress under the Debt Collection Improvement Act of 1996. See Section 2.11 for more information.

NPDES Program

- Federal civil penalties: Persons who discharge pollutants from a point source without a NPDES permit or in violation of that permit may be subject to the following: Administrative penalties up to $11,000 per day per violation; civil judicial penalties of up to $27,500 per day per violation.

- Federal criminal penalties: Penalties for negligent violation may include fines up to $25,000 per day of violation or one year imprisonment, or both. Penalties for knowing violations may include fines up to $50,000 per day of violation or imprisonment of up to three years, or both. Penalties for knowing endangerment may include fines up to $250,000 or imprisonment for not more than 15 years.

Oil Pollution Prevention Program

For discharges of oil or hazardous substances from onshore or offshore facilities, the owner, operator, or person in charge may be subject to the following:

- Federal civil penalties: Administrative penalties up to $11,000 per day per violation; civil judicial penalties of up to $27,500 per day of violation, or up to $1,100 per barrel of oil or unit of reportable quantity discharged. Violations which are the result of gross negligence or willful misconduct are subject to civil judicial penalties of not less that $110,000 and not more that $3,300 per barrel of oil or unit of reportable quantity discharged.

- Federal criminal penalties: Knowing violations may result in fines of up to $50,000 per day of violation, or up to three (3) years of imprisonment, or both; knowing endangerment may include fines up to $250,000 or imprisonment for not more than 15 years.

See Section 2.11.2 *Summary of Food Processing Cases in ECAARs from FY 1991 - 1997* for a description of CWA cases.

Note: EPA's Office of Water (202-260-5700) will direct callers with questions about the CWA to the appropriate EPA office. EPA also maintains a bibliographic database of Office of Water publications which can be accessed through the Ground Water and Drinking Water Resource Center at 202-260-7786.

2.3 Safe Drinking Water Act (SDWA)

The SDWA is the federal legislation that protects public health by regulating public drinking water and underground injection. EPA is responsible for writing regulations to carry out the provisions of the Act. Fifty-four of 56 states and territories have primacy to enforce compliance with National Primary Drinking Water Regulations (NPDWRs), as well as monitoring/reporting and public notification requirements contained in 40 CFR 141. EPA has primacy in Wyoming, Washington, D.C., and Tribal Lands, and may also take enforcement action in a primacy state where the state does not take an enforcement action in response to a violation. Generally speaking, most primacy states adopt drinking water regulations which closely reflect the federal requirements.

> *Safe drinking water regulations are discussed in more detail in Section 5.0 and Appendix A.2.*

EPA has developed national primary and secondary drinking water regulations under its SDWA authority, as well as monitoring/reporting, and public notification requirements. As part of the NPDWRs, EPA has developed maximum contaminant levels (MCLs) and treatment techniques (TTs) for more than 80 contaminants. MCLs are based on maximum contaminant level goals (MCLGs) and other factors. When there is no reliable method that is economically and technically feasible to measure a contaminant at particularly low concentrations, a TT is set rather than an MCL. Examples of TT rules are the Surface Water Treatment Rule and the Lead and Copper Rule. See Section 5.4.1 *National Primary Drinking Water Regulations* for more information.

National secondary drinking water regulations (NSDWRs) are federal guidelines regarding taste, odor, color, and certain other non-aesthetic effects of drinking water. These regulations are not federally enforceable. EPA recommends them to states as reasonable goals, but federal law does not require water systems to comply with them. States may however, adopt their own enforceable regulations governing these concerns. Therefore, check your state's drinking water regulations and contact your state regulatory agency.

In addition to EPA's SDWA requirements, water used in food processing operations must meet the Food and Drug Administration (FDA) and the United States Department of Agriculture (USDA) requirements. The FDA, under its good manufacturing practice regulations, requires that "any water that contacts food or food-contact surfaces shall be safe and of adequate sanitary quality" (Current Good Manufacturing Practice in Manufacturing, Packing, or Holding Human Food, 21 CFR 110.37). In addition, the USDA's Food Safety and Inspection Service (FSIS) sets standards for activities associated with the production of meat and poultry products, including standards involving water use and reuse.

For more information, see:

- Section 5.0 *How Do I Comply With Safe Drinking Water Regulations?*
- 40 CFR 141: National Primary Drinking Water Regulations
- 40 CFR 142: National Drinking Water Regulations Implementation
- 29 CFR 110: Current Good Manufacturing Practice in Manufacturing, Packing, or Holding Human Food
- •• Appendix A.2. *Summary of Principal Regulations Under the Safe Drinking Water Act.*

Underground Injection Control (UIC) Requirements. The SDWA UIC program (40 CFR 144-148) is a permit program designed to protect underground sources of drinking water by regulating the injection of liquid waste into five classes of injection wells. The UIC permit program is primarily enforced by primacy states; however, EPA maintains primacy for all wells in 13 states and territories, all Tribal Lands, and for some classes of wells in 7 states.

> *UIC requirements are discussed in Section 5.5 and Appendix A.2.*

If your facility meets certain criteria, you are required to obtain UIC authorization by permit or by rule to inject liquid waste. UIC permits include design, operating, inspection, closure, and monitoring requirements. Wells used to inject hazardous wastes also must comply with RCRA corrective action standards (40 CFR 264) in order to be granted a RCRA permit, and must meet applicable RCRA land disposal restriction (LDR) standards. See Section 8.6, Compliance Requirements for SQGs and LQGs, for more information on LDR standards.

For more information, see:

- Section 5.5 *Underground Injection Control (UIC) Requirements*
- 40 CFR 144-148: Underground Injection Control Program
- •• Appendix A.2: *Summary of Principal Regulations Under the Safe Drinking Water Act.*

Examples of SDWA Enforcement Provisions and Penalties

The 1986 amendments to the SDWA increased EPA's penalty authorities as an enforcement tool. Congress added new authority for assessment of administrative penalties and increased penalties for civil and criminal violations. These authorities were further strengthened for the Public Water Supply System (PWSS) program by the 1996 SDWA amendments. Some examples of EPA's enforcement authorities under the PWSS and UIC programs are summarized below.

PWSS Program

- Federal civil penalties: Persons who violate any applicable national primary drinking water regulation may be subject to the following penalties: up to $27,500 for failure to comply with any Administrative Compliance Order (any penalty sought in excess of $25,000, must be assessed by a civil judicial action); civil judicial penalties of up to $27,500 per day per violation.

- Federal criminal penalties: Persons who tamper with or attempt to tamper or threaten to tamper with a public water supply may be subject to the following: for tampering, not more than 5 years imprisonment, or fined in accordance with Title 18 U.S.C., or both; for attempting or threatening to tamper, not more than 3 years imprisonment, or fined in accordance with Title 18 U.S. C., or both; or a civil penalty of not more than $55,000 for any tampering, or not more than $22,000 for any attempt or threat.

Underground Injection Control

Persons who violate the requirements of an applicable UIC requirement may be subject to:

- Federal civil penalties: Administrative penalties up to $11,000 per day and civil judicial penalties of up to $27,500 per day per violation.

- Federal criminal penalties: Criminal penalties for willful violations may include fines in accordance with Title 18 U.S.C. or three years imprisonment, or both.

EPA also has emergency powers that are applicable to both PWSS and UIC when a contaminant, that may present an imminent and substantial endangerment to the health of persons, is present in or likely to enter a public water system or an underground source of drinking water.

> **Note: EPA's Safe Drinking Water Hotline (1-800-426-4791) provides answers to questions and distributes guidance pertaining to SDWA standards.**

2.4 Clean Air Act (CAA)

The Clean Air Act (CAA), including the Amendments (CAAA) of 1990, are designed to "protect and enhance the nation's air resources so as to promote the public health and welfare and the productive capacity of the population." Under the CAAA, many facilities will be required to obtain permits for the first time. State and local governments oversee, manage, and enforce many of the requirements of the CAAA.

> *CAA requirements are discussed in more detail in Section 6.0 and Appendix A.3.*

CAA Titles. The CAA consists of six sections, referred to as Titles, which direct EPA to establish national standards for ambient air quality. Titles I-VI regulations can be found in 40 CFR 50-95.

- **Title I - Air Pollution Prevention and Control.** Pursuant to Title I of the CAA, EPA has established national ambient air quality standards (NAAQSs) to limit levels of six **criteria pollutants**, including carbon monoxide, lead, nitrogen oxides, particulate matter (PM), ozone, and sulfur dioxide (40 CFR 50). Under Section 110 of the CAA, each state must develop a State Implementation Plan (SIP) to identify sources of air pollution and to determine what reductions are required to meet federal air quality standards. The SIP must be approved by EPA, or EPA may promulgate a plan of its own. Once a SIP is approved, it may be enforced by both federal and state authorities (CAA Section 110, 42 U.S.C., Section 7410(a)(2)). Geographic areas that meet NAAQSs for a given pollutant are classified as attainment areas; those that do not meet NAAQSs are classified as nonattainment areas. Those areas that are classified as nonattainment must update their SIPs in order to improve air quality.

Title I also authorizes EPA to establish New Source Performance Standards (NSPSs), which are nationally uniform emission standards for new stationary sources falling within particular industrial categories (CAA Section 111). NSPSs are based on the pollution control technology available to that category of industrial source but allow the affected industries the flexibility to devise a cost-effective means of reducing emissions.

- **Title II - Emission Standards for Moving Sources.** Title II of the CAA (Section 201-250) pertains to mobile sources, such as cars, trucks, buses, and planes. It establishes allowable levels of automobile emissions and includes provisions for alternative fuels. Reformulated gasoline, automobile pollution control devices, and vapor recovery nozzles on gas pumps are a few of the mechanisms EPA uses to regulate mobile air emission sources.

- **Title III - Air Toxics.** Under Title I, EPA establishes and enforces National Emissions Standards for Hazardous Air Pollutants (NESHAPs), nationally uniform standards oriented towards controlling particular hazardous air pollutants (HAPs). Title III further directed EPA to develop a list of sources that emit any of 188 HAPs listed under Section 112 of the CAA, and to develop regulations for these categories of sources. To date, EPA has listed 174 categories and developed a schedule for the establishment of emission standards. The emission standards will be developed for both new and existing sources based on maximum achievable control technology (MACT). The MACT is defined as the control technology achieving the maximum degree of reduction in the emission of the HAPs, taking into account cost and other factors.

- **Title IV - Acid Deposition Control.** Acid rain occurs when sulfur dioxide and nitrogen oxide emissions are released into the atmosphere and return to the earth in rain, fog, or snow. Title IV establishes a sulfur dioxide emissions program designed to reduce the formation of acid rain by requiring power plants and other utilities to reduce sulfur dioxide emissions. Reduction of sulfur dioxide releases will be obtained by granting certain sources limited emissions allowances. This program began in 1995 and set levels of sulfur dioxide releases below previous levels.

- **Title V - Permits.** Title V of the CAAA of 1990 created a permit program for all **major sources** (and certain other sources) regulated under the CAA. One purpose of the operating permit is to include, in one document, all air emissions requirements that apply to a given facility. States are developing the permit programs in accordance with guidance and regulations from EPA. Once a state program is approved by EPA, permits will be issued and monitored by that state.

- **Title VI - Stratospheric Ozone.** Title VI is intended to protect stratospheric ozone by phasing out the manufacture of ozone-depleting chemicals and restricting their use and distribution. Title VI requires EPA to list all regulated substances along with their ozone depletion potential, atmospheric lifetimes, and global warming potentials. Production of Class I substances, including 15 kinds of chlorofluorocarbons (CFCs), will be phased out entirely by the year 2000, while certain hydrochlorofluorocarbons

(HCFCs) will be phased out by 2030. Title VI also requires EPA to publish a list of substitutes for Class I and II chemicals.

Risk Management Program. As required under Section 112(r) of the amended CAA, EPA has promulgated the Risk Management Program Rule. The rule's main goals are to **prevent accidental releases** of regulated substances and to reduce the severity of those releases that do occur by requiring facilities to develop risk management programs. The risk management programs must incorporate three

> *Food processors may be subject to risk management planning requirements if they have one or more of the identified substances onsite above the threshold quantity. Ammonia is one of the identified substances.*

elements: a hazard assessment, a prevention program, and an emergency response program. These programs are to be summarized in a risk management plan (RMP) that will be made available to state and local government agencies and the public. Besides helping facilities prevent accidents, the rule can improve the efficiency of work operations by ensuring that workers are trained in proper procedures and by using preventive maintenance to reduce equipment breakdowns.

If you have more than a threshold quantity of any of the **regulated substances** in a single process, you are required to comply with the regulation (40 CFR 68). EPA has currently established a list of 140 regulated substances that fall under these CAA regulations. These substances were published in the *Federal Register* on January 31, 1994; EPA amended the list by rule, published on December 18, 1997. EPA may amend the list in the future as needed. Covered facilities must comply with the rule **by June 21, 1999.** The RMPs will be available electronically to state and local governments and citizens to help them understand local chemical hazards and take steps to prevent accidents.

For more information on risk management planning, contacting the RCRA/UST, Superfund and EPCRA Hotline at 1-800-424-9346 or 703-412-9810, or access EPA's Chemical Emergency Preparedness and Prevention Office Home Page at **http://www.epa.gov/swercepp/.**

For more information, see:

- Section 6.0 *How Do I Comply With Air Regulations?*
- 40 CFR 50-99: Air Programs
- 40 CFR 68: Chemical Accident Prevention Provisions
- •• Appendix A.3. *Summary of Principal Regulations Under the Clean Air Act.*

Examples of CAA Enforcement Provisions and Penalties

The 1990 Clean Air Act Amendments gave EPA additional enforcement authorities including administrative authorities and field-issued citations. Field-issued citations are those that are issued by a compliance inspector at the time of an inspection. Civil penalty amounts presented here also reflect the inflation adjustment authorized by Congress under the Debt Collection Improvement Act of 1996 (see Section 2.11 for more information).

- Federal civil penalties: Failure to comply with an air operating permit, the State Implementation Plan (SIP) or a federal regulation may result in a civil judicial penalty of up to $27,500 per day per violation. A field-issued citation may result in penalties up to $5,000 per violation.

- Federal criminal penalties: Knowing violation may result in criminal penalties including fines up to $1,000,000 per day per violation and fifteen years imprisonment.

See Section 2.11.2 *Summary of Food Processing Cases in ECAARs from FY 1991 - 1997* for a description of CAA cases.

Note: EPA's Control Technology Center (919-541-0800) provides general assistance and information on CAA standards. The Stratospheric Ozone Information Hotline (1-800-296-1996) provides general information about regulations promulgated under Title VI of the CAA, and EPA's RCRA/UST, Superfund and EPCRA Hotline (1-800-424-9346 or 703-412-9810) provides information concerning accidental release prevention under CAA Section 112(r). In addition, the Technology Transfer Network Bulletin Board System (modem access 919-541-5742) includes recent CAA rules, EPA guidance documents, and updates of EPA activities.

2.5 Emergency Planning And Community Right-To-Know Act (EPCRA)

The Superfund Amendments and Reauthorization Act (SARA) of 1986 created EPCRA, also known as SARA Title III. This statute was designed to improve community access to information about chemical hazards and to facilitate the development of chemical emergency response plans by state and local governments. EPCRA required the establishment of state emergency response commissions (SERCs). SERCs are responsible for coordinating certain emergency response activities and for appointing local emergency planning committees (LEPCs).

> *EPCRA requirements are discussed in more detail in Section 7.0 and Appendix A.4. CERCLA requirements are discussed in more detail in Section 10.0 and Appendix A.5.*

EPCRA regulations establish four types of reporting obligations for facilities which store or manage specified chemicals:

- ***EPCRA Section 302/303.*** Requires facilities to notify the SERC (EPCRA Section 302) and LEPC (EPCRA Section 303) of the presence of any extremely hazardous substance (the list of such substances in 40 CFR 355, Appendices A and B) if it has such a substance in excess of the substance's threshold planning quantity.

- ***EPCRA Section 304.*** Requires the facility to notify the SERC and the LEPC in the event of an accidental release exceeding the reportable quantity of an EPCRA extremely hazardous substance or a CERCLA hazardous substance. Facilities are also required to notify the National Response Center at 1-800-424-8802 in the event of a release of a CERCLA hazardous substance.

- ***EPCRA Sections 311 and 312.*** Require a facility at which a hazardous chemical [as defined by the Occupational Safety and Health Administration (OSHA)] or an EPCRA extremely hazardous substance is present in an amount exceeding a specified threshold to submit to the SERC, LEPC, and local fire department material safety data sheets (MSDSs) or lists of MSDSs and hazardous chemical inventory forms (also known as Tier I and II forms). This information helps the local government respond in the event of a spill or release of the chemical.

- ***EPCRA Section 313.*** Commonly referred to as the Toxic Chemical Release Inventory (TRI), this program requires certain designated businesses to submit annual reports (known as Form Rs and Form As) on more than 600 EPCRA Section 313 chemicals and chemical categories. Facilities meeting the EPCRA Section 313 reporting criteria must report the annual releases and other waste management activities (routine and accidental) of EPCRA Section 313 chemicals to all environmental media. The reports are submitted to U.S. EPA and State or Tribal governments, on or before July1, for activities in the previous calendar year. This information increases the public's knowledge or, and access to information on the presence of toxic chemicals in their communities.

All information submitted pursuant to EPCRA regulations is publicly accessible, unless protected by a trade secret claim.

For more information, see:

- Section 7.0 *How Do I Comply With the Emergency Planning and Community Right-to-Know Act Requirements?*
- 40 CFR 350-372: Emergency Planning and Community Right-to-Know Information
- Appendix A.4. *Summary of Principal Regulations Under the Emergency Planning and Community Right-to-Know Act.*

Examples of EPCRA Enforcement Provisions and Penalties

Examples of civil and criminal penalties under EPCRA are described below. Civil penalty amounts presented here also reflect the inflation adjustment authorized by Congress under the Debt Collection Improvement Act of 1996. (See Section 2.11 for more information.)

- Federal civil penalties: Failure to do the following may result in civil penalties of up to $27,500 per day per violation: submit Forms Rs for all EPCRA 313 chemicals for which the facility exceeded the threshold; provide information in response to a request from the local emergency planning committee; report accidental releases to all appropriate authorities; provide all reporting information required; or notify the committee of any changes at the facility that affect the development of the emergency response plan.

- Federal criminal penalties: Knowing violation may result in criminal penalties including fines up to $25,000 and/or up to two years in prison.

See Section 2.11.2 *Summary of Food Processing Cases in ECAARs from FY 1991 - 1997* for a description of EPCRA cases.

Note: EPA's RCRA/UST, Superfund and EPCRA Hotline (1-800-424-9346 or 703-412-9810) provides information and distributes guidance regarding the emergency planning and community right-to-know regulations.

2.6 Resource Conservation and Recovery Act (RCRA)

The Resource Conservation and Recovery Act (RCRA) of 1976 which amended the Solid Waste Disposal Act, addresses solid (Subtitle D) and hazardous (Subtitle C) waste management activities. The Hazardous and Solid Waste Amendments (HSWA) of 1984 strengthened RCRA's waste management provisions and added Subtitle I, which governs underground storage tanks (USTs).

Hazardous waste requirements are discussed in more detail in Section 8.0 and Appendix A.6.

Subtitle D of RCRA and its implementing regulations basically apply to the management of solid, nonhazardous waste and its disposal in landfills. Subtitle D applies to your food processing facility because it prohibits open dumping of solid, nonhazardous wastes. A nonhazardous waste is defined as any garbage, refuse, or sludge from waste treatment plants, water treatment plants, or air pollution control equipment. Programs addressing the disposal of solid, nonhazardous wastes are developed and enforced at the state or local level. Contact your state for more information on proper disposal practices.

Regulations promulgated pursuant to **Subtitle C** of RCRA (40 CFR 260-299) establish a "cradle-to-grave" system governing hazardous waste from the point of generation to disposal. RCRA hazardous wastes include the specific materials listed in the regulations or materials which exhibit a hazardous waste characteristic (ignitability, corrosivity, reactivity, or toxicity).

Regulated entities that generate hazardous waste are subject to waste accumulation, manifesting, and recordkeeping standards. Facilities that treat, store, or dispose of hazardous waste must obtain a permit, either from EPA or from a state agency which EPA has authorized to implement the permitting program. Subtitle C permits contain general facility standards such as contingency plans, emergency procedures, recordkeeping and reporting requirements, financial assurance mechanisms, and unit-specific standards. RCRA also contains provisions (40 CFR 264 Subpart S and 264.10) for conducting corrective actions which govern the cleanup of releases of hazardous waste or constituents from solid waste management units at RCRA-regulated facilities.

Most RCRA requirements are not industry specific but apply to any company that generates, transports, treats, stores, or disposes of hazardous waste. Although RCRA is a federal statute, many states implement the RCRA program. Currently, EPA has delegated its authority to 46 states to implement various provisions of RCRA. UST programs are delegated to about half of the states. Important RCRA regulatory requirements include:

- **Identification of Solid and Hazardous Wastes** (40 CFR 261) lays out the procedure every generator should follow to determine whether the material created is considered a solid waste, hazardous waste, or is exempt from regulation.

> *Food processors typically generate small amounts of hazardous waste. See Section 8.0 for more information.*

- **Standards for Generators of Hazardous Waste** (40 CFR 262) establishes the responsibilities of hazardous waste generators including obtaining an identification (ID) number, preparing a manifest, ensuring proper packaging and labeling, meeting standards for waste accumulation units, and recordkeeping and reporting requirements. Generators can accumulate hazardous waste for up to 90 days (or 180 days depending on the amount of waste generated per month) without obtaining a permit for being a treatment, storage, and disposal (TSD) facility.

- **Land Disposal Restrictions (LDRs)** (40 CFR 268) are regulations prohibiting the disposal of hazardous waste on land without prior treatment. Under the LDRs, materials must meet LDR treatment standards for hazardous constituents prior to placement in a RCRA land disposal unit (landfill, land treatment unit, waste pile, or surface impoundment). Land disposal units are defined in 40 CFR 264 and 265, Subparts K-N. Generators of waste subject to the LDRs must provide notification of such to the designated TSD facility to ensure proper treatment prior to disposal.

- **Used Oil Management Standards** (40 CFR 279) impose management requirements affecting the storage, transportation, burning, processing, and re-refining of the used oil. For parties that merely generate used oil, regulations establish storage standards. For a party considered a used oil marketer (one who generates and sells off-specification used oil directly to a used oil burner), additional tracking and paperwork requirements must be satisfied, including registration form EPA 8700-12.

- **Containers** (40 CFR 264 and 265, Subpart I; 40 CFR 261.7) are one of the most commonly used and diverse forms of hazardous waste storage. There are two sets of regulations for containers: requirements that pertain to the management of hazardous waste containers (40 CFR 264/265, Subpart I) and the regulations governing residues of hazardous waste in empty containers (40 CFR 261.7).

- **Tanks** (40 CFR 265, Subpart J) are used widely for storage or accumulation of hazardous waste because they can accommodate huge volumes. Generators accumulating hazardous waste in tank systems are subject to the interim status provisions in 40 CFR 265, Subpart J.

- **Emissions - Tanks and Containers** (40 CFR 264 and 265, Subpart CC) used to store hazardous waste with a high volatile organic concentration must meet emission standards under RCRA. Regulations require generators to test the waste to determine the concentration of the waste, to satisfy tank and container emissions standards, and to inspect and monitor regulated units. These regulations apply to all facilities that store such waste, including generators operating under the 90-day accumulation rule.

- **Storage Tanks - USTs** (40 CFR 280) containing petroleum and hazardous substances are regulated under RCRA, Subtitle I. Subtitle I regulations contain tank design and release detection requirements, as well as financial responsibility and corrective action standards for USTs. The UST program also establishes standards for upgrading existing tanks, that must be met by December 22, 1998.

  Note: Aboveground storage tanks (ASTs) may be used to store materials, such as vegetable oils, used in food processing. ASTs are regulated under the CWA and OPA. Refer to Section 4.0 *How Do I Comply with Wastewater Discharge and Related Regulations?* for more information on ASTs.

For more information, see:

- Section 8.0 *How Do I Comply With the Hazardous Waste Regulations?*
- 40 CFR 260-299: Hazardous Waste Management
- •• Appendix A.6. *Summary of Principal Regulations Under the Resource Conservation and Recovery Act.*

Examples of RCRA Enforcement Provisions and Penalties

General enforcement penalty information is presented below for the solid waste, hazardous waste, and underground storage tank categories of RCRA regulations. Civil penalty amounts presented here also reflect the inflation adjustment authorized by Congress under the Debt Collection Improvement Act of 1996 (see Section 2.11 for more information).

Solid Waste

- Federal law does not establish specific penalties for civil or criminal violations of the solid waste program. Enforcement of the solid waste program relies on state law.

Hazardous Waste

Federal law for the hazardous waste management program has provisions for civil and criminal penalties.

- Federal civil penalties: Civil penalties may be up to $27,500 per day of noncompliance per incident and the company's permit may be revoked.

- Federal criminal penalties: The criminal penalties apply to individuals within a company and are a maximum of two years (five years for specified violations) imprisonment and a maximum penalty of $50,000 per day of noncompliance per incident.

Underground Storage Tanks

Failure to comply with UST requirements may result in the following types of civil penalties:

- Federal civil penalties: Administrative penalties may be up to $11,000 per violation per tank per day of noncompliance. Failure to comply with an enforcement order can result in civil judicial penalties of up to $27,500 per day of noncompliance with the order.

See Section 2.11.2 *Summary of Food Processing Cases in ECAARs from FY 1991 - 1997* for a description of RCRA cases.

> **Note: EPA's RCRA/UST, Superfund and EPCRA Hotline (1-800-424-9346 or 703-412-9810) provides information and distributes guidance regarding all RCRA regulations.**

2.7 Comprehensive Environmental Response, Compensation, And Liability Act (CERCLA)

The Comprehensive Environmental Response, Compensation, and Liability Act (CERCLA), a 1980 law commonly known as Superfund, authorizes EPA to respond to releases, or threatened releases, of hazardous substances that may endanger public health, welfare, or the environment. SARA revised various sections of CERCLA, extended the taxing authority for Superfund and creating a free-standing law, SARA Title III, also known as EPCRA (discussed in Section 2.5).

Release Reporting. When there is a release of a CERCLA hazardous substance in an amount equal to or in excess of a certain quantity for that substance, CERCLA requires the person in charge of a vessel or facility to immediately notify the **National Response Center at 1-800-424-8802** (40 CFR 302, CERCLA 103(a)). See Section 7.0 for more information on EPCRA and CERCLA emergency release reporting requirements.

> *CERCLA release reporting requirements are discussed in more detail in Section 7.0, 10.0, and Appendix A.5.*

Responses. EPA implements hazardous substance responses, known as remedial actions or removals, according to procedures outlined in the National Oil and Hazardous Substances Pollution Contingency Plan (NCP) (40 CFR 300). While EPA generally takes remedial actions only at National Priorities List (NPL) sites, both EPA and states can act at other sites. The difference is that EPA can provide responsible parties the opportunity to conduct removal and remedial actions and encourages community involvement throughout the Superfund response process.

For more information, see:

- Section 7.0 *How Do I Comply With the Emergency Planning and Community Right-to-Know Act Regulations?*

- Section 10.0 *Other Major Environmental Statutes and Regulations: CERCLA, RCRA Subtitle D, FIFRA and TSCA*
- 40 CFR 300: National Oil and Hazardous Substances Pollution Contingency Plan
- 40 CFR 302: Hazardous Substance Release Reporting Regulations
- Appendix A.5. *Summary of Principal Regulations Under the Comprehensive Environmental Response, Compensation, and Liability Act*

Examples of CERCLA Enforcement Provisions and Penalties

Civil penalty amounts presented here also reflect the inflation adjustment authorized by Congress under the Debt Collection Improvement Act of 1996 (see Section 2.11 for more information).

- Federal civil penalties: Civil penalties of up to $27,500 per day per violation for the first violation and a second violation can be as high as $82,000 per day.

- Federal criminal penalties: Persons in charge of a facility from which a hazardous substance is released and who violate CERCLA's notification requirements (i.e., fail to notify the required government agency or knowingly submit false information) are subject to penalties under Title 18 or imprisonment for up to three years, or both. Persons who knowingly fail to notify EPA of hazardous substance management activities are subject to penalties up to $10,000 or imprisonment for up to one year.

> **Note: EPA's RCRA/UST, Superfund and EPCRA Hotline (1-800-424-9346 or 703-412-9810) provides information and references guidance pertaining to the Superfund program.**

2.8 Toxic Substances Control Act (TSCA)

Under TSCA, EPA collects data on chemicals in order to evaluate, assess, mitigate, and control risks which may be posed by their manufacture, processing, and use. TSCA provides a variety of control methods to prevent chemicals from posing unreasonable risk, and the standards may apply at any point during a chemical's life cycle. Drugs, cosmetics, foods, food additives, pesticides, and nuclear materials are **exempt from TSCA** and are subject to control under other federal statutes (e.g., foods and food additives are under the purview of the Federal Food, Drug and Cosmetics Act (FFDCA) administered by the FDA. In order for a food or food additive to be exempt, however, it must meet the definition contained in the FFDCA (21 USC 321 et seq.), or related statutes such as the Poultry Products Inspection Act and the Federal Meat Inspection Act. If the food or food additive does not meet the definition, the substance may then be regulated under TSCA and is subject to all the requirements of TSCA including testing, premanufacture notice, reporting and recordkeeping, export notification, and import certification. For example, vegetable oils and their derivatives from vegetable processing that are used as an ingredient in lubricants, paints, inks, fuels, plastics, solvents and a variety of other industrial products are subject to all of TSCA's requirements.

> *TSCA requirements are discussed in more detail in Section 10.4.*

Section 8 of TSCA authorizes EPA to require chemical manufacturers, importers, and processors to keep records and report certain information. This includes reporting as part of the inventory update (Section 8(a)); maintaining and reporting allegations of signification adverse reactions (Section 8(c)); reporting health and safety studies (Section 8(d)); and reporting information on a substances presenting a substantial risk of injury to health or the environment (Section 8(e). Additional reporting requirements for exports and imports are found in TSCA Sections 12 and 13, respectively.

The TSCA Chemical Substances Inventory is a compilation of the names of all existing chemical substances and currently contains over 70,000 existing chemicals. Information in the inventory is updated every four years (Inventory Update). If manufacturing or importing a

> *Food processors manufacturing substances, such as vegetable oil and animal fats, that are used for non-food purposes (e.g., in inks) must comply with the Inventory Update rule.*

chemical substance that is not already on the inventory (and has not been excluded by TSCA), a facility must submit a premanufacture notice (PMN) prior to manufacture or importation (TSCA Section 5).

For more information, see:

- Section 10.0 *Other Major Environmental Statutes and Regulations: CERCLA, RCRA Subtitle D, FIFRA and TSCA*
- 40 CFR 704: Reporting and Recordkeeping Requirements
- 40 CFR 707: Chemical Imports and Exports
- 40 CFR 710: TSCA Chemical Inventory
- 40 CFR 712: Chemical Information Rules
- 40 CFR 716: Health and Safety Data Reporting
- 40 CFR 717: Records and Reports of Allegations that Chemical Substances Cause Significant Adverse Reactions to Health or the Environment
- 40 CFR 720: Premanufacture Notice
- 40 CFR 723: Premanufacture Notification Exemptions
- 40 CFR 721: Significant New Uses of Chemical Substances
- 40 CFR 750: Procedures for Rulemaking Under Section 6 of TSCA
- 40 CFR 790: Test Rule Development and Exemption Procedures
- 40 CFR 791: Data Reimbursement
- 40 CFR 792: Good Laboratory Practice Standards.

Examples of TSCA Enforcement Provisions and Penalties

TSCA Section 11 gives EPA broad authority to inspect establishments which hold chemicals and to subpoena information for enforcement of the Act. TSCA Sections 15, 16, and 17, respectively, list prohibited acts, their attendant civil and criminal penalties, and the jurisdiction of the federal district court for specific enforcement and seizure. Civil penalty amounts presented here also reflect the inflation adjustment authorized by Congress under the Debt Collection Improvement Act of 1996 (see Section 2.11 for more information).

•• Federal civil penalties: Civil penalties of up to $27,500 per day per violation.

• Federal criminal penalties: Criminal penalties may be up to $25,000 per day per violation and/or imprisonment for up to one year.

Note: EPA's TSCA Assistance Information Service (202-554-1404) provides information and distributes guidance pertaining to TSCA standards.

2.9 Federal Insecticide, Fungicide, and Rodenticide Act (FIFRA)

The Federal Insecticide, Fungicide, and Rodenticide Act (FIFRA) primarily regulates the manufacture and registration of pesticides (40 CFR 152 and 156), but important requirements also exist for pesticides **users**.

Pesticide handling requirements are discussed in more detail in Section 10.3.

FIFRA requires that all pesticides be registered for every intended use, and that labels containing instructions for proper storage, use, and disposal accompany each pesticide marketed. It is considered illegal to use a pesticide in a manner inconsistent with its label. The "label is the law." Under FIFRA, pesticides must be classified for either general use or restricted use. EPA classifies some pesticides as restricted use because they have high toxicity or pose particular environmental hazards. Restricted use pesticides may be applied only by certified pesticide applicators. Pesticide labels will state clearly whether a particular pesticide is restricted use only. For pesticides that are not restricted use, food processing facilities may purchase, store, apply, and dispose of the pesticides. Food processors must comply with all FIFRA requirements relating to these activities.

Food Quality Protection Act

The Food Quality Protection Act (FQPA), passed in 1996, was a comprehensive overhaul of the laws that regulate pesticides in food: FIFRA and the Federal Food, Drug and Cosmetics Act (FFDCA). The new law amends both major pesticide laws to establish a more consistent, protective regulatory scheme. The

FQPA requirements are discussed in more detail in Section 10.3.

new **FFDCA provisions** include establishing a health-based safety standard for pesticide residues in food; adding special provisions for infants and children; placing limitations on benefits considerations; reviewing all existing tolerances within ten years; incorporating endocrine testing; enhancing enforcement of pesticide residue standards by allowing the FDA to impose civil penalties for tolerance violations; increasing right to know activities; and requiring uniformity of tolerances among states (unless the state petitions EPA for an exception, based on state-specific situations). The new **FIFRA provisions** include a pesticide reregistration program, pesticide registration renewal, registration of safer pesticides, minor use pesticide program, and an antimicrobial pesticide program.

For more information, see:

- Section 10.0 *Other Major Environmental Statutes and Regulations: CERCLA, RCRA Subtitle D, FIFRA and TSCA*
- 40 CFR 150: FIFRA
- 40 CFR 165: Regulations for the Acceptance of Certain Pesticides and Recommended Procedures for the Disposal and Storage of Pesticides and Pesticides Containers.

Examples of FIFRA Enforcement Provisions and Penalties

FIFRA. Civil penalty amounts presented here also reflect the inflation adjustment authorized by Congress under the Debt Collection Improvement Act of 1996 (see Section 2.11 for more information).

- Federal civil penalties: Commercial applicators may be fined up to $5,500 for each offense under FIFRA; private applicators may be fined $550 for the first offense and up to $1,000 for each subsequent offense.

- Federal criminal penalties: Commercial applicators may be fined up to $25,000 or up to one year in prison, or both, for knowing violations. Private applicators may be fined up to $1,000 or thirty days in prison, or both, for knowing violations.

> **Note: EPA's National Pesticides Telecommunications Network (NPTN) at 1-800-858-7378 provides pesticide information.**

2.10   Other Federal Regulations

This subsection highlights two other environmental laws that may affect food processors, particularly in construction projects for new facilities or modifications of existing facilities. These include the Coastal Zone Management Act (CZMA) and the Endangered Species Act (ESA). You should be aware of and comply with the requirements of these regulations as described below. For purposes of this guide, additional information about these statutes is incorporated in Section 4.7.2. *Construction and Plant Modification Activities*.

2.10.1   Coastal Zone Management Act (CZMA)

The CZMA, enacted in 1972 and administered by the National Oceanic and Atmospheric Administration (NOAA), encourages states to preserve, protect, develop, and where possible, restore or enhance valuable natural coastal resources such as wetland, floodplains, estuaries, beaches, dunes, barrier islands, and coral reefs, as well as the fish and wildlife using those habitats.

A unique feature of the CZMA is that participation by states is voluntary. The CZMA enables states to develop programs and plans that meet their specific needs, within the context of their governmental structures. In addition, CZMA gives states the authority to review federal projects and projects receiving federal licenses and permits to ensure they abide by state laws, regulations, and policies. To encourage states to participate, the act makes federal financial assistance available to any coastal state or territory, including those on the Great Lakes, that is willing to develop and implement a comprehensive coastal management program (CMP). In addition to resource protection, the CZMA specifies that coastal states may manage coastal development. A state with an approved program can deny or restrict any development that is inconsistent with its CMP.

Under the 1990 CZMA Reauthorization Amendments, states must issue management measures for certain categories of runoff and erosion; evaluate nonpoint sources; and identify coastal areas that would be affected negatively by specified land uses. The 1990 Amendments mandate each coastal state to implement a Coastal Zone Nonpoint Pollution Control Program as part of each state's CMP. For example, under the program, pesticide application is subject to regulation if pesticide runoff from nonpoint sources reaches coastal waters.

> *States may add additional requirements to NPDES storm water permits in order to meet coastal zone nonpoint pollution control program goals. See Section 4.3 for more information.*

Consequently, food processors who use pesticides and live in coastal states should determine whether their land is part of the coastal zone, or if their pesticide application violates their state's applicable CMP.

The CZMA also was amended by the Coastal Zone Protection Act of 1996. This act amends the CZMA to change allowable uses and match requirements for certain grant funds and to change the process for consistency appeals.

For more information, see:

- Section 4.3 *Am I A Direct Discharger?*
- 16 U.S.C. Sections 1451-1464
- 15 CFR 921-932.

2.10.2 Endangered Species Act (ESA)

The Endangered Species Act (ESA), administered by the U.S. Department of Interior's Fish and Wildlife Service (USFWS) and the Department of Commerce's National Marine Fisheries Service (NMFS), establishes a program for the conservation of endangered and threatened species and the habitats in which they are found. The ESA affords broad protection for species of fish, wildlife, and plants that are listed as endangered and threatened in the U.S. and elsewhere. Provisions are made for listing species, as well as for recovery plans and the designation of critical habitat for listed

> *ESA requirements that may affect food processors are discussed in more detail in Section 4.7.2.*

species. Anyone can petition the USFWS to list a species. The ESA strives to conserve ecosystems both through federal action and by encouraging the establishment of state programs. State laws or regulations may be more, but not less, restrictive than the federal ESA or its regulations.

> The term "take" includes harassing, harming, hunting, killing, capturing, and collecting.

The ESA prohibits the taking, possession, import, export, sale, and transport of any listed fish or wildlife species. It also is unlawful to maliciously damage, destroy, or remove from any area under federal jurisdiction, damage or remove from any other area in knowing violation of state law, import, export, or trade any listed plant species. These prohibitions do not apply to species legally held in captivity or a controlled environment. The USFWS or NMFS, by permit, also may allow a taking incidental to an otherwise lawful activity if the applicant submits, and the USFWS or NMFS approves, a conservation plan addressing the impact of the taking, mitigation measures, funding, and alternative actions considered.

Persons engaged in, or planning to engage in, activities such as construction or plant modification, must be aware if any endangered or threatened species exist on the property involved or if the property is considered part of a listed species' critical habitat. If neither is the case, the ESA does not apply. However, if the action will "take" a species or degrade critical habitat, some form of mitigating action must be taken to prevent harming the species.

For more information, see:

- Section 4.7.2 *Construction and Plant Modification Activities*
- 16 U.S.C. Sections 1531-1544
- 50 CFR 10, 13, 14, 17, and 23.

2.11 Summary Of The Enforcement Process and Selected Cases

2.11.1 Overview of Enforcement

Some of the statute-specific enforcement authorities that Congress gave EPA are described earlier in this section. To provide a context for those examples, the following briefly describes the roles of EPA and the states in environmental enforcement, particularly under delegated or approved state programs, and the general types of enforcement responses available to EPA. Citizen suit authority also is discussed briefly.

> Environmental enforcement is a comprehensive program involving federal, state, and local governments.

Federal Government - Roles of EPA and DOJ

EPA leads the federal government's environmental enforcement efforts using the latest law enforcement techniques and drawing upon the specialized abilities of other federal agencies. EPA headquarters, located in Washington, D.C., includes the Office of Enforcement and Compliance Assurance (OECA) which manages the Agency's enforcement and litigation program. Ten EPA regional offices, located in cities such as Seattle, Atlanta, San Francisco, and Philadelphia, conduct most of the day-to-day enforcement activities of the Agency. Where a state has been approved by EPA to implement a program, the EPA regional office oversees the state's performance to assure consistency with the federal law (see below). In unapproved states, the EPA regional office administers the program.

OECA includes the National Enforcement Investigations Center (NEIC) in Denver, and a Criminal Investigation Division (CID), headquartered in Washington, D.C., with field offices in the ten EPA regions as well as other locations around the country. NEIC provides technical support for EPA's civil and criminal cases. CID is the only federal law enforcement agency created for the purpose of investigating environmental crimes, although environmental crimes sometimes are investigated by the FBI and other federal agencies.

The U.S. Department of Justice (DOJ) plays a crucial role in EPA's enforcement activities. When EPA wishes to prosecute a violator in the U.S. court system, EPA refers the case to DOJ. DOJ attorneys, who specialize in environmental litigation, consider EPA's recommendations and make the final decision of whether or not to file the case in federal court. When the case goes to court, DOJ represents EPA in court, though EPA's legal and technical staff remain actively involved in the case. Like EPA, DOJ has a field organization -- the U.S. Attorneys; however, its civil environmental cases are handled by mostly DOJ headquarters attorneys.

State Government - Definition of Delegated or Approved Programs

Virtually every federal environmental law allows state governments to develop programs for implementing the federal law in their states. When a state submits a complete application and EPA has determined that the state program meets the federal requirements, EPA approves the state program. Depending on which federal statute, such programs are called "delegated," "authorized," "approved" or "primacy" programs. After EPA approves a program, the state applies the national standards and regulations by issuing and enforcing its own rules and permits. Many of EPA's statutes allow Native American Tribal Governments to develop programs for implementing the federal laws on Tribal lands, by means similar to EPA's process for delegating programs to states. As a matter of policy, EPA has extended this option to the other statutes that do not explicitly provide for delegation to Native American Tribal Governments. Hence, the potential exists to delegate to Native American Tribal Governments any program that EPA may delegate to states. In practice however, the number of Tribal Governments with delegated responsibilities is small. If you do operate a food processing facility on Tribal lands, you should check with the EPA's Regional Office and/or the Tribal Government to learn whether EPA has approved any Tribal Government environmental programs.

Under this system of delegated or approved programs, state governments carry out the vast majority of environmental enforcement actions. State governments conduct about 80-90 percent of the inspections and approximately 70 percent of the national total of the enforcement actions taken by the delegated clean air, clean water, and hazardous waste programs.

Enforcement at the state level is carried out by a number of different agencies, including the state environmental agencies, state Attorney General, and district attorneys. State environmental agencies usually have responsibility for permits, inspections, and certain types of enforcement actions. In many instances, pesticide laws are enforced by state Departments of Agriculture. In some states, a single environmental agency handles all EPA programs, while in others, several agencies divide the responsibilities. States also may delegate some of the activities to county or city governments, such as the local health department.

The state Attorney General is the chief law enforcement official for the state. The Attorney General has responsibility for suing violators, at the request of state environmental agencies. District attorneys also may have responsibility for suing violators, and typically they represent municipalities. In some states, the District Attorney's approval is required for enforcement suits to be filed by the state Attorney General.

State/Federal Enforcement Partnerships

EPA strives to work out an effective enforcement partnership with each state. This is accomplished by establishing state/EPA enforcement partnership agreements that cover delegated programs and involve the appropriate state agencies. These agreements usually define the characteristics of a good enforcement program using the same criteria by which EPA judges its own performance.

The agreements also spell out the conditions under which EPA will step in and take enforcement action in a delegated or an approved state program. Common circumstances for such a step include the following:

- If the state requests federal action;
- If the state's enforcement response is not timely and appropriate (a set of criteria has been developed by EPA and the states for each major program);
- If the case involves national precedents; or
- If there is a violation of an EPA order or consent decree (settlement agreement).

Types of Enforcement Responses

Enforcement actions are tools designed to discourage companies and individuals in the regulated community from breaking the law, and to compel them to **return to compliance** when they do break the law. EPA has a range of options when contemplating an enforcement response against a violator. These options differ from one law to another, and include the following:

- *Informal response.* Administrative actions that are advisory in nature, such as a phone call, notice of noncompliance or a warning letter. In these actions, EPA advises the manager of a facility what violation was found, what corrective action should be taken, and by what date. Informal responses carry no penalty or power to compel actions, but if they are ignored, they can lead to more severe actions.

- *Formal administrative responses.* Legal orders that are independently enforceable and which may require the recipient to take some corrective or remedial action within a specified period of time, refrain from certain behavior, or be in future compliance. (Such an order may or may not have a penalty attached.) These administrative actions are strong enforcement tools. If a person violates an order, EPA may go to U.S. federal court to force compliance. Administrative actions are handled under EPA's internal administrative litigation system, which is comparable to any court system except that administrative law judges (ALJs) preside.

- **Civil judicial responses.** Formal lawsuits brought in U.S. federal court by DOJ at EPA's request. They normally are used against the more serious or recalcitrant violators of environmental laws or to seek prompt correction of imminent hazards. Civil judicial cases generally result in penalties and court orders requiring correction of the violation and specific actions to prevent future violations.

- **Criminal judicial responses.** Response used when a person or company has knowingly violated the law. In a criminal case, DOJ prosecutes an alleged violator in federal court, seeking criminal sanctions including fines and imprisonment. Criminal actions are often used to respond to flagrant, intentional disregard for environmental laws (such as operating secret by-pass pipes to discharge untreated wastewater and deliberate falsification of reports or records).

In many enforcement actions, EPA seeks both a remedy and a penalty. The remedy includes returning the violating facility to compliance and sometimes other remedial actions, as described below.

- **Compliance.** The violator will be required to comply with the law. If the violation has not already been corrected, the violator usually is placed under a court-ordered schedule, with severe penalties for failure to comply with the order.

- **Benefit projects.** In some cases, the violator is permitted to carry out a supplemental environmental project (SEP) that will yield environmental benefits. These projects may partly offset the penalty and may mitigate the harmful effects of the violation.

- **Penalties.** The violator is required to pay a cash penalty (in criminal cases, a fine) that is not tax deductible. The penalty includes sanctions intended to deter the violator from falling into noncompliance again and to deter others from similar violations. [1]

- **Imprisonment.** In criminal cases, the violator may be sentenced to jail or placed on probation.

- **Contractor listing.** A facility that has violated the CWA or CAA may be placed on EPA's List of Violating Facilities. Listed facilities are not eligible to receive federal contracts, grants, or loans from EPA or any other federal agency. Facilities that commit criminal violations of other environmental statutes are subject to possible

---

[1] **Civil Monetary Penalty Inflation Adjustment Rule.** This new rule (January 30, 1997) and the associated policy modified all of EPA's existing civil penalty policies by increasing the **gravity** component for civil monetary penalties by ten percent. EPA's action was based on the Debt Collection Improvement Act (DCIA) of 1996 that Congress enacted to restore the deterrent effect of federal civil penalties, eroded by inflation over the years. The law requires each federal agency to adjust its civil monetary penalties in accordance with a specified formula. EPA is required to review and adjust these amounts every four years. EPA's *Civil Monetary Penalty Inflation Adjustment Rule,* codified in 40 CFR 19, *Adjustment of Civil Penalties for Inflation,* increased **all 65** of the Agency's civil penalty provisions (with the exception of the 1996 Safe Drinking Water Act penalty provisions) by ten percent -- the maximum that Congress allowed for the first adjustment due to inflation. EPA's Office of Enforcement and Compliance Assurance (OECA) also issued a new penalty policy, *Modification to EPA Penalty Policies to Implement the Civil Monetary Penalty Inflation Rule* (May 9, 1997) See 40 CFR 19.4, Table 1, for a complete list of all EPA's civil monetary penalty authorities and amounts, or see the 1997 policy and related materials on OECA's home page at **http//www.epa.gov/oeca/.**

suspension and/or debarment from receiving or entering into EPA or other federal agency contracts.

Citizen Suit Provisions

The first citizen suit provision appeared in 1970, when Congress enacted the CAA. Specifically, this provision allowed citizens to sue polluters who violated certain requirements of the CAA and to sue EPA if it failed to carry out a non-discretionary duty set forth in the Act. Since that time, Congress has incorporated citizen suit provisions into many, but not all, federal environmental statutes. Although these provisions vary from statute to statute, such provisions generally allow citizen groups or individuals to file actions in federal district court against a facility to correct violations or collect fines and penalties.

2.11.2 Summary of Food Processing Cases in ECAARs from FY 1991-1997

For the past several years, EPA has published annual reports, the *Enforcement and Compliance Assurance Accomplishments Reports (ECAARs)*, on the accomplishments of the environmental enforcement and compliance assurance program. Although the organization of these reports has changed over the years, each report contains narrative descriptions of significant administrative, civil judicial and criminal cases that were either taken, developed, and/or settled by EPA and the states.

Most of the cases in each report reflect those that have been concluded by some type of settlement agreement, either administrative or civil judicial, or by court order. In a few instances the same case may appear in both an earlier and a later report, as it moves from the stage of being filed to being concluded. The conclusion might be a consent agreement that was negotiated over more than one year or a court order following a trial. In a criminal case, the sentencing of a convicted defendant(s) may be reported in the next year's report. Because the same case may appear in more than one report, a small amount of double counting results.

This summary is based on **78 cases** selected from all cases described in the ECAARs for fiscal years (FYs)1991-1997. These cases were chosen on the basis of the facility name, description of the type of business operation, or, in some instances, on the listed SIC Code. It's important to note that the cases described in each report **do not necessarily reflect all** of the cases affecting food processors in that particular fiscal year.

CWA Cases. More than one third of all the 78 ECAAR-reported cases over the seven year period involved violations under the CWA. Examples of violations include the following: exceeding NPDES discharge limits (BOD, TSS, temperature, pH, phosphorus, oil and grease); exceeding indirect discharge limits (BOD, ammonia); interference and pass through at a POTW; and illegal discharges to surface waters (beer, ammonia, blood wastes; groundwater contaminated with solvents). These cases resulted in civil penalties ranging from a low of $14,000 to a high of $12.6 million.

Several cases involved criminal acts, including the following: conspiracy to violate the CWA; falsifying discharge monitoring reports (DMRs) sent to EPA or the state; negligently or knowingly discharging pollutants without a permit; operating a secret by-pass that resulted in discharge of untreated wastewater; and other acts. These cases resulted in criminal fines for the companies and/or the individuals involved. In addition, convicted individuals were sentenced either to incarceration in federal prison followed by a term of supervised release, or to a combination of in-home incarceration and community service.

CAA Cases. About one sixth of the 78 ECAAR-reported cases in the seven year period involved violations of the CAA. Examples of violations include: exceeding limits on boiler emissions (particulates); opacity; exceeding limits on volatile organic compound (VOC) emissions (ethanol); asbestos demolition and removal; prevention of significant deterioration (PSD) violations such as constructing of a major source without a permit; and violations of NSPS requirements. These cases resulted in civil penalties ranging from a low of $30,000 (opacity violations) to a high of $385,000 (VOC violations).

One criminal case involved illegal removal and release of asbestos to the air and resulted in a $350,000 fine for a food processor.

EPCRA Cases. Slightly less than one third of the 78 ECAAR-reported cases involved violations of EPCRA. On average, two cases were reported each fiscal year, until the FY 1997 EPCRA Section 312 Food Processing Sector Initiative which resulted in ten cases. Examples of violations include: failure to submit TRI Form Rs (ammonia, sulfuric acid, hydrochloric acid, and/or carbon dioxide); failure to submit material safety data sheets (MSDSs) to LEPCs; failure to submit Tier I/Tier II forms; and failure to report emergency releases of anhydrous ammonia to state and local authorities. Failure to report these same release to EPA was a violation of CERCLA. Therefore, several companies had violations under both EPCRA and CERCLA.

Penalties in the EPCRA cases ranged from a low of $2,000 (under the FY 1997 EPCRA Section 312 Food Processing Sector Initiative) to a high of about $73,000. The penalties in the combined EPCRA/CERCLA cases ranged from a low of $41,000 to a high of $180,830.

RCRA Cases. Only four of the 78 ECAAR-reported cases in the seven year period involved RCRA. Examples of violations include: violation of the used oil requirements; failure to make a hazardous waste determination; and accumulating hazardous waste onsite in excess of 90 days. Penalties in these cases ranged from $250,000 to $700,000.

NOTICE

This document provides guidance to assist regulated entities to understand their obligations under environmental laws; however, for a complete understanding of all legal requirements, the reader must refer to applicable federal and state statutes and regulations. This guide is a compliance assistance tool only, and it neither changes nor replaces any applicable legal requirements, nor does it create any rights or benefits for anyone. This guide also describes in a summary fashion the roles and activities of federal and state agencies; however, the guidance does not limit their otherwise lawful prerogatives, and the agencies may act at variance with it, based on specific circumstances. This guidance may be revised without prior notice. Mention of trade names or commercial products in this document, or in associated references, does not constitute an endorsement or recommendation for use.

PREFACE AND ACKNOWLEDGMENTS

Preface

As part of its mission to communicate environmental regulatory responsibilities to business and industry, the U.S. Environmental Protection Agency's Office of Compliance (OC) has prepared this guide to the major federal environmental statutes and regulations that may affect food processors. The guide provides an overview of the major requirements that the U.S. EPA administers. Appendices A, B and C, respectively, contain portions of the Code of Federal Regulations (CFR) in an easy to understand format, provide organizations and hotline resources for compliance assistance, and list numerous references used in developing this guide. The guide also contains general information about pollution prevention that may enable your facility to go beyond compliance by achieving greater reductions in emissions and/or wastes.

The target audience for the guide is the plant-level manager and/or staff responsible for environmental compliance at a facility. Others who may find value in this guide include the following: environmental managers at the corporate level; state and local compliance assistance programs; trade associations; and environmental consultants to the industry. Federal, state, and local regulators and/or compliance inspectors may also find this guide to be useful in offering a broad perspective on U.S. EPA's requirements and the food processing industry.

Acknowledgments

The Chemical, Commercial Services and Municipal Division (CCSMD) of OC received assistance from the following groups in planning and developing this guide. Representatives of the American Frozen Food Institute (AFFI), the American Meat Institute (AMI) and the National Food Processors Association (NFPA) supported this project, and offered informed judgments about its structure and contents. The Food Industry Environmental Council (FIEC), an organization of trade associations and companies, provided valuable assistance by securing plant-level review and comment on an early draft, and by coordinating review of the draft final guide among its entire membership.

Project Manager: Ms. Rebecca A. (Becky) Barclay   Contract Support: SAIC, Dunn Loring, VA
U.S. EPA, Office of Compliance   Contract #: 68-C4-0072
Washington, D.C.   Work Assignment#: EC-3-7 (OC)
E-mail address: barclay.rebecca@epamail.epa.gov   WAM: Ms. Joletta Humpert,
Telephone: (202) 564-7063   Environmental Scientist
CCSMD Web site: http://es.epa.gov/oeca/ccsmd/.

TABLE OF CONTENTS

Table of Contents (continued)

Table of Contents (continued)

Table of Contents (continued)

LIST OF TABLES

LIST OF FIGURES

ACRONYMS

ABC	Activity-Based Costing
ACP	Area Contingency Plan
AFFI	American Frozen Food Institute
AFO	Animal Feeding Operation
AHERA	Asbestos Hazard Emergency Response Act
ALJ	Administrative Law Judge
AMI	American Meat Institute
AO	Administrative Order
AST	Aboveground Storage Tank
BACT	Best Available Control Technology
BIFs	Boilers and Industrial Furnaces
BOD	Biochemical Oxygen Demand
BMP	Best Management Practice
CAA	Clean Air Act
CAAA	Clean Air Act Amendments
CAM	Compliance Assurance Monitoring
CCSMB	Chemical, Commercial Services, and Municipal Branch
CDC	Center for Disease Control
CEPPO	Chemical Emergency Preparedness and Prevention Office
CERCLA	Comprehensive Environmental Response, Compensation, and Liability Act
CESQG	Conditionally Exempt Small Quantity Generator
CFC	Chlorofluorocarbon
CFR	Code of Federal Regulation
CH_4	Methane
CID	Criminal Investigation Division
CMP	Coastal Management Program
CO	Carbon Monoxide
COD	Chemical Oxygen Demand
CWA	Clean Water Act
CZMA	Coastal Zone Management Act
DMR	Discharge Monitoring Report
DOE	Department of Energy
DOI	Department of the Interior
DOJ	Department of Justice
DOT	Department of Transportation
ECAAR	Enforcement and Compliance Assurance Accomplishments Report
ECOS	Environmental Council of the States
EHS	Extremely Hazardous Substance
ELP	Environmental Leadership Program
EMS	Environmental Management System
EPCRA	Emergency Planning and Community Right-to-Know Act
EPA	U.S. Environmental Protection Agency
ERNS	Emergency Response Notification System
ESA	Endangered Species Act
FBI	Federal Bureau of Investigations
FDA	Food and Drug Administration

FEPCA	Federal Environmental Pesticide Control Act
FESOP	Federally Enforceable State Operating Permit
FFDCA	Federal Food, Drug and Cosmetic Act
FIEC	Food Industry Environmental Council
FIFRA	Federal Insecticide, Fungicide, and Rodenticide Act
FMC	Food Manufacturing Coalition
FQPA	Food Quality Protection Act
FRP	Facility Response Plan
FSIS	Food Safety Inspection Service
FWPCA	Federal Water Pollution Control Act
FY	Fiscal Year
HACCP	Hazard Analysis and Critical Control Point
HAP	Hazardous Air Pollutant
HAZWOPER	Hazardous Waste Operations and Emergency Response
HCFC	Hydrochlorofluorocarbon
HCS	Hazard Communication Standard
HSWA	Hazardous Solid Waste Amendments
ICCR	Industrial Combustion Coordinated Rulemaking
ID	Identification Number
ISO	International Organization of Standardization
LAER	Lowest Achievable Emission Rate
LDR	Land Disposal Restriction
LEPC	Local Emergency Planning Committee
LQG	Large Quantity Generator
MACT	Maximum Achievable Control Technology
MCL	Maximum Contaminant Level
MCLG	Maximum Contaminant Level Goal
MEK	Methyl Ethyl Ketone
MSDS	Material Safety Data Sheet
MSGP	Multi-Sector General Permit
N_2O	Nitrous Oxide
NAAQS	National Ambient Air Quality Standard
NAICS	North American Industrial Classification System
NCP	National Oil and Hazardous Substances Pollution Contingency Plan
NEIC	National Enforcement Investigations Center
NESHAP	National Emissions Standards for Hazardous Air Pollutants
NFPA	National Food Processors Association
NICE[3]	National Industrial Competitiveness through Energy, Environment, and Economics
NMFS	National Marine Fisheries Service
NOx	Nitrogen Oxide
NO_2	Nitrogen Dioxide
NOAA	National Oceanic and Atmospheric Administration
NOI	Notice of Inspection
NPDES	National Pollutant Discharge Elimination System
NPDWR	National Primary Drinking Water Regulation
NPL	National Priority List
NPTN	Pesticides Telecommunications Network
NRC	National Response Center
NSDWR	National Secondary Drinking Water Regulation
NSPS	New Source Performance Standard

NSR	New Source Review
NTIS	National Technical Information Service
OC	Office of Compliance
OECA	Office of Environmental Compliance and Assurance
O&G	Oil and Grease
OPA	Oil Pollution Act
OPC	Oil Program Center
OPP	Office of Pesticide Programs
OPPT	Office of Prevention, Pesticides, and Toxics
OSC	On-Scene Coordinator
OSHA	Occupational Safety and Health Administration
OTAG	Ozone Transport Assessment Group
P2	Pollution Prevention
Pb	Lead
PCB	Polychlorinated biphenyl
PCS	Permit Compliance System
PESP	Pesticide Environmental Stewardship Program
PM	Particulate Matter
PMN	Premanufacture Notice
POTW	Publicly Owned Treatment Work
PREP	National Preparedness for Response Exercise Program
PSD	Prevention of Significant Deterioration
PSM	Process Safety Management
PTI	Permit to Install
PWS	Public Water System
PWSS	Public Water Supply Supervision
RA	Regional Administrator
RACT	Reasonably Available Control Technology
RCRA	Resource Conservation and Recovery Act
RMP	Risk Management Plan
RQ	Reportable Quantity
RUP	Restricted Use Pesticide
SARA	Superfund Amendments and Reauthorization Act
SDWA	Safe Drinking Water Act
SEP	Supplemental Environmental Project
SERC	State Emergency Response Commission
SIC	Standard Industrial Classification
SIP	State Implementation Plan
SNAP	Significant New Alternatives Policy
SOx	Sulfur Oxide
SO$_2$	Sulfur Dioxide
SPCC	Spill Prevention, Control and Countermeasure
SQG	Small Quantity Generator
SRF	State Revolving Fund
SWPPP	Storm Water Pollution Prevention Plan
SWTR	Surface Water Treatment Rule
tpy	Tons per year
TCLP	Toxicity Characteristic Leaching Procedure
TKN	Total Kjeldahl Nitrogen
TOC	Total Organic Carbon
TPQ	Threshold Planning Quantity

TRI	Toxic Release Inventory
TSCA	Toxic Substances Control Act
TSD	Treatment, Storage, and Disposal facility
TSS	Total Suspended Solids
TT	Treatment Technique
TTN	Technology Transfer Network
UIC	Underground Injection Control
USCG	U.S. Coast Guard
USDA	U.S. Department of Agriculture
USDHHS	U.S. Department of Human and Health Services
USDW	Underground Source of Drinking Water
USFWS	U.S. Fish and Wildlife Service
UST	Underground Storage Tank
VOC	Volatile Organic Compound
WPS	Worker Protection Standard

SECTION 3 CONTENTS

3. UNDERSTANDING THE PROCESS: INPUTS, OUTPUTS, AND APPLICABLE FEDERAL ENVIRONMENTAL REGULATIONS

3.1 Introduction

The section provides you with an approach for analyzing your facility's operations to identify the wastes generated and how those wastes are regulated.

Remember that this guide discusses the most significant, but not all, of the federal environmental requirements that apply to your food processing facility. State and local requirements are not addressed.

First, this section leads you through an examination of the activities at a typical food processing facility, including process and ancillary operations. It will (1) describe the inputs and the waste outputs generated during process and ancillary operations, and (2) identify the federal environmental requirements associated with the waste outputs. To help you visualize the steps, this section includes figures (generic Figures 3-1a and 3-1b) that show typical process and ancillary operations for the food processing industry, and their inputs and regulated waste outputs.

After reviewing this generic model of a food processing operation, the next example will show you a process map (Figure 3-2) for a facility in Standard Industrial Classification (SIC) Code 203, including typical inputs, regulated outputs, and the applicable environmental statute.

The final part of this section provides you with an opportunity to examine your facility's process and ancillary operations, identify inputs and waste outputs, and determine how they are regulated. A blank waste analysis process map (Figure 3-3) and a blank waste analysis table (Table 3-3) are provided to help you in this activity.

3.2 Examining Process and Ancillary Operations

The process map of your food processing operation, as well as your ancillary operations, are most likely very similar to those shown in the following figures:

Figure 3-1a. Generic Process Map With Examples of Regulated Outputs
Figure 3-1b. Selected Ancillary Operations with Examples of Regulated Outputs.

As shown in these figures, you will find that your process and ancillary operations are comprised of various inputs and associated outputs of waste. Inputs, which can range from raw ingredients to hazardous materials (see Section 3.2.1), and waste outputs vary greatly depending on type(s) of food products being produced. The applicable environmental statute for each type of waste output is indicated in parentheses on generic Figures 3-1a and 3-1b. Sections 3.2.1 and 3.2.2, respectively, discuss inputs and outputs in greater detail.

3.2.1   Inputs

As shown in generic **Figures 3-1a and 3-1b**, inputs go into every step of the process and ancillary operations. Inputs can consist of a variety of materials, including raw products, chemicals, water, paper, ink, steam, etc. The inputs to each operation will vary depending on the type of facility and product(s) being produced.

Hazardous Materials. To meet your input needs, your food processing facility may store and use many types of hazardous or toxic materials in your daily operations including, but not limited to, oils, chemicals, paints, pesticides, and fuels. Many of these materials may be regulated because of their hazardous or toxic nature. Please note that the term "materials" is not an EPA regulatory term, but a broad term selected for purposes of this discussion.

EPA and other federal regulations use various terms to denote hazardous or toxic materials. Examples of several terms used to denote these types of materials include the following:

- EPA refers to regulated materials by terms such as "hazardous substances" and "extremely hazardous substances" under the Emergency Planning and Community Right-to-Know Act (EPCRA) and the Comprehensive Environmental Response, Compensation, and Liability Act (CERCLA). Lists of these substances can be found in the EPCRA/CERCLA regulations at 40 CFR 302, Table 302.4 and 355, respectively. Such regulated materials do not have to be waste outputs to be covered under these regulations. In fact, such materials may be inputs to your process or ancillary operations.

- A "hazardous material" is defined by the U.S. Department of Transportation (DOT) as a substance or material...capable of posing an unreasonable risk to health, safety, and property when transported in commerce, and which has been so designated. For DOT, this term includes hazardous and extremely hazardous substances as defined in CERCLA/EPCRA, hazardous wastes as defined in Resource Conservation and Recovery Act (RCRA), marine pollutants, and elevated temperature materials.

For some types of hazardous or toxic materials, EPA regulates storage and how you are to report your use of them. The typical practices for storage and handling of hazardous materials are designed to prevent the following: exposure to individuals, releases to the environment, and mixing (which could cause explosions, fire, or unwanted chemical reactions and releases). See Section 9.0 *How Do I Comply With Spill or Chemical Release Requirements?* for a multimedia overview of requirements, or, for more specific detail, see each statute-specific section (Sections 4.0 through 10.0).

Figure 3-1a. Generic Process Map with Examples of Regulated Outputs

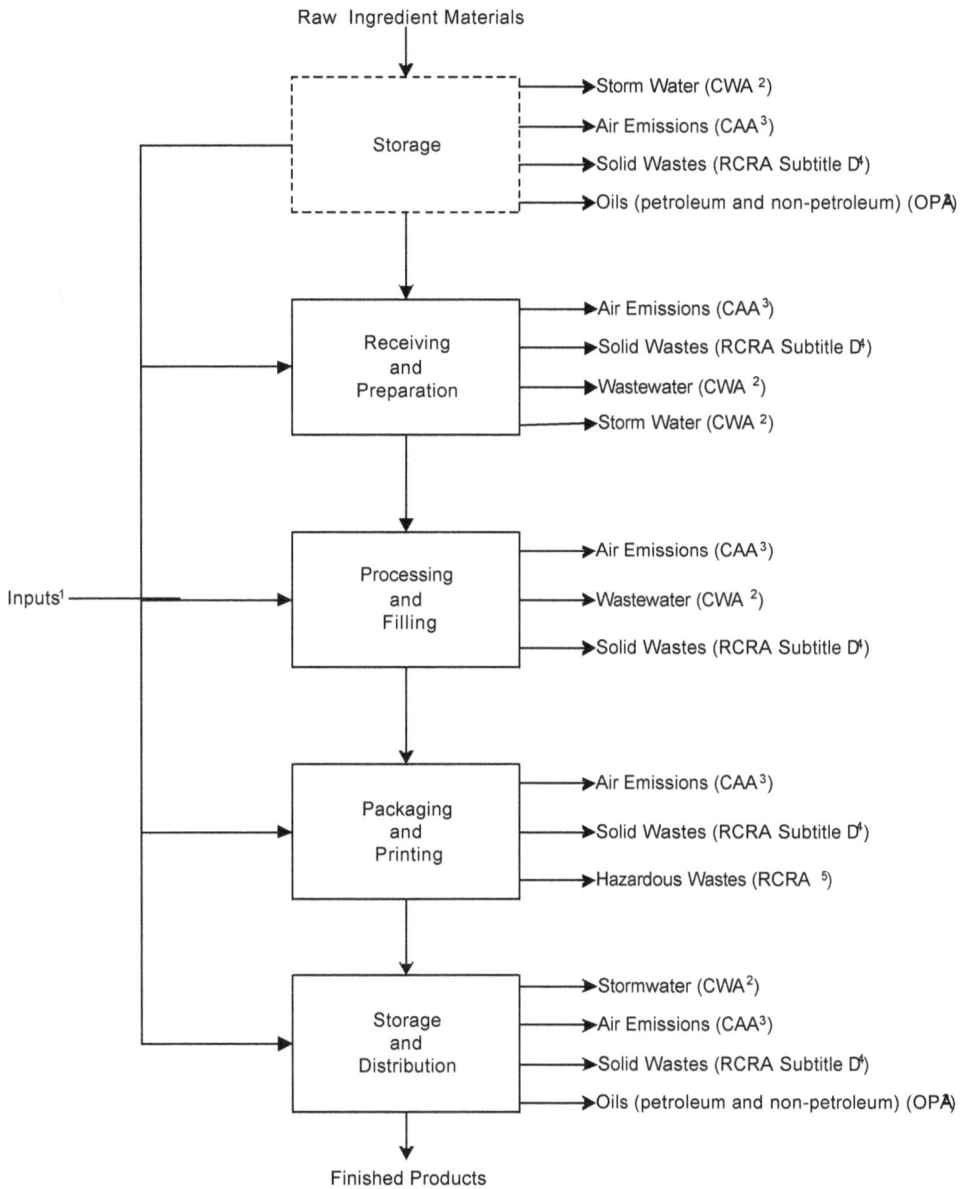

[1] Some Inputs may be regulated under federal statutes (e.g., CERCLA, EPCRA).
See Section 3.2 (following) and Section 7.0 How Do I Comply with the Emergency Planning and Community Right-to-Know Act Regulations?
[2] See Section 4.0 How Do I Comply with Wastewater Discharge and Related Regulations?
[3] See Section 6.0 How Do I Comply with Air Regulations?
[4] See Section 10.0 Other Major Environmental Statutes and Regulations: CERCLA, RCRA Subtitle D, FIFRA, and TSCA.
[5] See Section 8.0   How Do I Comply with the Hazardous Waste Regulations?

Figure 3-1b. Selected Ancillary Operations with Examples of Regulated Outputs

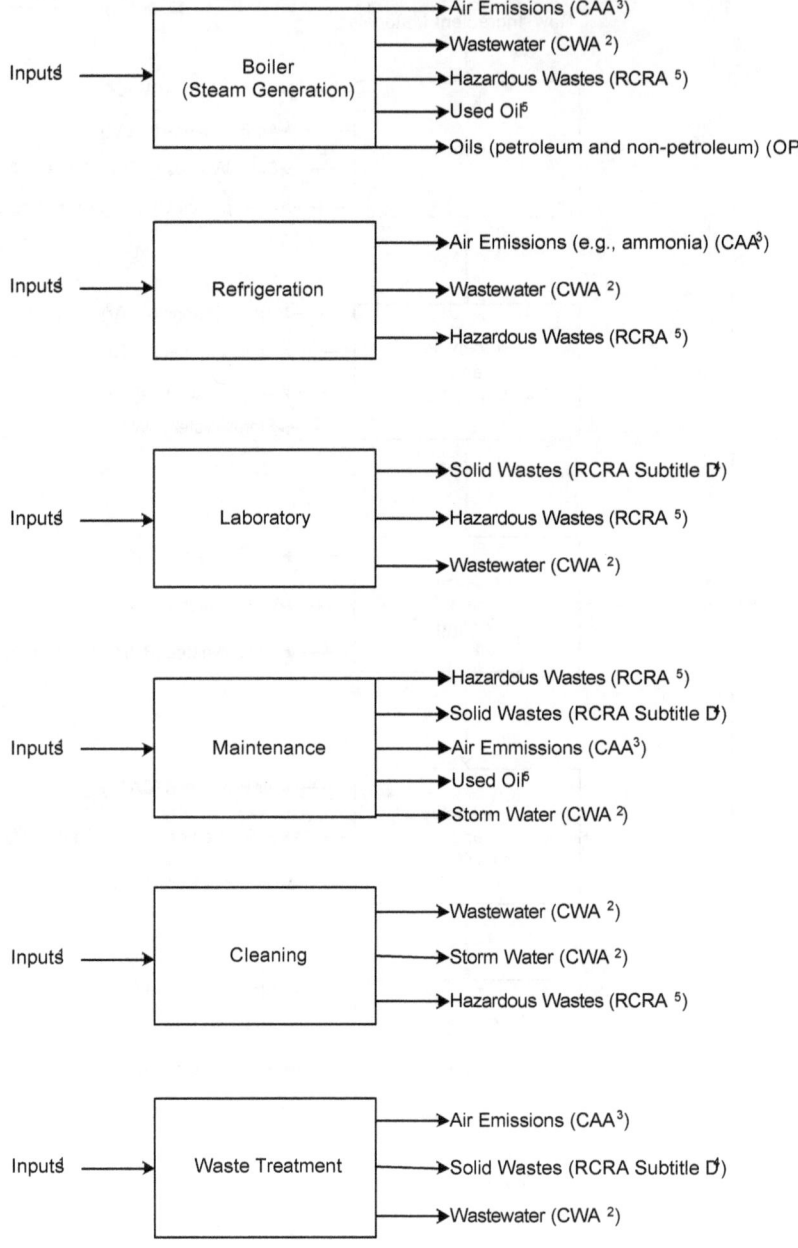

[1] Some Inputs may be regulated under federal statutes (e.g., CERCLA, EPCRA).
See Section 3.2 (following) and Section 7.0 How Do I Comply with the Emergency Planning and Community Right-to-Know Act Regulations?
[2] See Section 4.0 How Do I Comply with Wastewater Discharge and Related Regulations?
[3] See Section 6.0 How Do I Comply with Air Regulations?
[4] See Section 10.0 Other Major Environmental Statutes and Regulations: CERCLA, RCRA Subtitle D, FIFRA, and TSCA.
[5] See Section 8.0 How Do I Comply with the Hazardous Waste Regulations?

It is important to understand the difference between hazardous or toxic materials and hazardous wastes. For the purposes of this discussion, the term hazardous or toxic materials includes all materials that have **not been used, and therefore, are not wastes**. Thus, hazardous materials include those hazardous and extremely hazardous substances as defined in CERCLA/EPCRA, but not hazardous wastes as defined by RCRA.

Hazardous Wastes. Hazardous wastes are those materials which are no longer usable and are to be disposed of. Hazardous wastes must be managed according to the RCRA hazardous waste regulations. See Section 8.0 *How Do I Comply With the Hazardous Waste Regulations?* The RCRA regulations also address non-hazardous wastes (e.g., solid wastes). See Section 10.2 *Subtitle D of the Resource Conservation and Recovery Act* for more information.

Example of Hazardous Material Versus Hazardous Waste: Methyl Ethyl Ketone (MEK)

Hazardous Material   A drum of MEK being **stored** at a facility is a **hazardous material** under EPCRA regulations. It is not a classified as a **hazardous waste** under RCRA because <u>it is not a waste</u>.

Hazardous Waste   As a waste, MEK is a **RCRA-listed hazardous waste**. A drum of MEK that cannot be used (e.g., is contaminated during use, exceeds its shelf-life, or is off-spec), becomes a waste and must be disposed of as a hazardous waste. Additionally, MEK that is released in the event of a spill or accidental release, must be managed as a hazardous waste.

3.2.2   Overview of Outputs and Applicable Statutes

Outputs from food processing include the saleable products being produced and the wastes. This section will focus on the wastes and the environmental regulations that apply to their management and/or disposal. You must first identify all the wastes your facility generates.

Wastes are generated throughout your process (see generic Figure 3-1a), as well as from your ancillary operations (see generic Figure 3-1b). Many activities occur during each part of the process that generate wastes, including the following:

Process Step	Activity Generating Waste
Storage	Storage of raw materials, refrigeration, and onsite transport.
Receiving and Preparation	Loading, conveyor handling, cleaning, inspection, sorting, separating, washing, peeling, cutting, and pulverizing. Also includes water unloading and fluming.
Processing and Filling	Mixing, cooking, freezing, concentrating, freeze-drying, filling, cooling, preserving, and flavoring.

Process Step	Activity Generating Waste
Packaging and Printing	Can-making, printing, packaging (e.g., plastic bag, paper, can, glass jars and bottles, cardboard, and pallet-packaging).
Storage and Distribution	Storage of prepared materials, refrigeration, and loading.

The wastes generated can take one of the four forms called *wastestreams*, including wastewaters, air emissions, hazardous wastes, and solid wastes. Each of these wastestreams is regulated by one or more environmental statutes as follows:

- *Wastewater* is regulated under the Clean Water Act (CWA). Additionally, some discharges of wastewater to underground injection wells are regulated under the Safe Drinking Water Act (SDWA).

- *Air emissions* are regulated under the Clean Air Act (CAA). Some air emissions, such as those from waste storage or the burning of hazardous waste, are regulated under RCRA.

- *Hazardous wastes* are regulated under RCRA.

- • Solid wastes are regulated under RCRA Subtitle D.

In addition to the regulatory background information provided in Section 2.0 *Guide to EPA's Major Environmental Statutes*, the following sections, organized by statute (with the exception of Section 9.0), provide additional information on regulatory compliance requirements your facility must follow when managing these wastestreams:

- Section 4.0 How Do I Comply with Wastewater Discharge and Related Regulations?

- Section 5.0 How Do I Comply with Safe Drinking Water Regulations?

- Section 6.0   How Do I Comply with Air Regulations?

- Section 7.0   How Do I Comply with the Emergency Planning and Community Right-To-Know Act Regulations?

- Section 8.0   How Do I Comply with the Hazardous Waste Regulations?

- Section 9.0   How Do I Comply With Spill or Chemical Release Requirements? For purposes of comparison, this section pulls together and briefly summarizes your responsibilities for emergency planning and response requirements across several statutes [e.g., EPCRA, CERCLA, CWA, Oil Pollution Act (OPA), CAA, RCRA]. Always refer to the statute-specific section, the regulations, or program guidance for additional information.

•• Section 10.0   Other Major Environmental Statutes and Regulations: CERCLA, RCRA Subtitle D, FIFRA, and TSCA.

3.3   Conducting a Waste Analysis

3.3.1   Example Waste Analysis For SIC 203 Facility

The following discussion presents an example (in **Figure 3-2** and **Table 3-2**) of how to apply this method of examining process inputs and outputs (e.g., wastes) and identifying applicable environmental requirements for a facility in SIC Code 203, *Canned, Frozen and Preserved Fruits, Vegetables and Food Specialties*. The types of facilities included in SIC Code 203 are presented in **Table 3-1**.

Table 3-1. Types of SIC 203 Facilities

SIC	NAICS *	Types of Facilities
2032	311422 and 311999	Canned Specialties
2033	311421	Canned Fruits, Vegetables, Preserves, Jams and Jellies
2034	311423 and 311211	Dried and Dehydrated Fruits, Vegetables, and Soup Mixes
2035	311421 and 311941	Pickled Fruits and Vegetables, Vegetable Sauces and Seasonings, and Salad Dressings
2037	311411	Frozen Fruits, Fruit Juices, and Vegetables
2038	311412	Frozen Specialties, Not Elsewhere Classified

*   The 1997 North American Industrial Classification System (NAICS) codes for the food processing industry will replace the 1987 SIC codes in publications of the U.S. Statistical Agencies over several years (1998-2004), beginning with publications of the NAICS United States Manual. The NAICS Implementation Schedule for these agencies is available on the U.S. Census Bureau's Internet site at **http://www.census.gov/epcd/naics/timeschd.html/**.

Figure 3-2 presents a waste analysis for a hypothetical SIC Code 203 facility. The common process activities include: (1) storage (e.g., storage of raw produce); (2) receiving and preparation (e.g., sorting fruits and vegetables to remove immature or substandard ones; cleaning and washing; peeling [sometimes using caustic solutions to remove peels]; and coring and pitting); (3) processing and filling; (4) packaging and printing; and (5) storage and distribution. Typical inputs and outputs are identified for each process activity.

Figure 3-2. Process Waste Analysis for a SIC 203 Facility

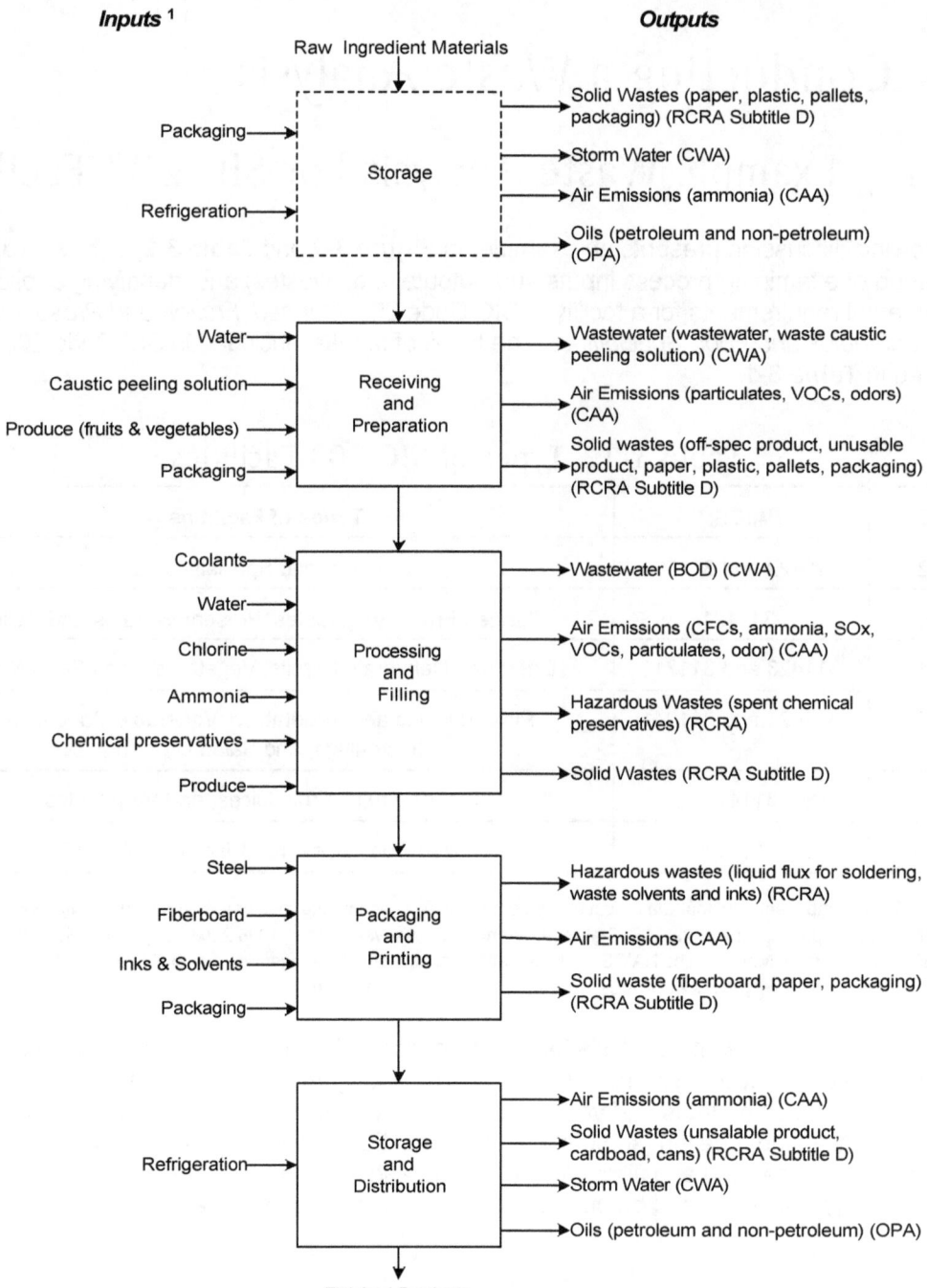

Inputs [1]

Outputs

Raw Ingredient Materials

Packaging →

Refrigeration →

Storage

→ Solid Wastes (paper, plastic, pallets, packaging) (RCRA Subtitle D)

→ Storm Water (CWA)

→ Air Emissions (ammonia) (CAA)

→ Oils (petroleum and non-petroleum) (OPA)

Water →

Caustic peeling solution →

Produce (fruits & vegetables) →

Packaging →

Receiving and Preparation

→ Wastewater (wastewater, waste caustic peeling solution) (CWA)

→ Air Emissions (particulates, VOCs, odors) (CAA)

→ Solid wastes (off-spec product, unusable product, paper, plastic, pallets, packaging) (RCRA Subtitle D)

Coolants →

Water →

Chlorine →

Ammonia →

Chemical preservatives →

Produce →

Processing and Filling

→ Wastewater (BOD) (CWA)

→ Air Emissions (CFCs, ammonia, SOx, VOCs, particulates, odor) (CAA)

→ Hazardous Wastes (spent chemical preservatives) (RCRA)

→ Solid Wastes (RCRA Subtitle D)

Steel →

Fiberboard →

Inks & Solvents →

Packaging →

Packaging and Printing

→ Hazardous wastes (liquid flux for soldering, waste solvents and inks) (RCRA)

→ Air Emissions (CAA)

→ Solid waste (fiberboard, paper, packaging) (RCRA Subtitle D)

Refrigeration →

Storage and Distribution

→ Air Emissions (ammonia) (CAA)

→ Solid Wastes (unsalable product, cardboad, cans) (RCRA Subtitle D)

→ Storm Water (CWA)

→ Oils (petroleum and non-petroleum) (OPA)

Finished Products

[1] Some, not all, of the inputs listed in this figure may be regulated under federal statutes (e.g., CERCLA, EPCRA). See Section 3.2 and Section 7.0 How Do I Comply With the Emergency Planning and Community Right-to-Know Act Regulations?

To further explore this facility's wastes, **Table 3-2** presents examples of typical wastes from each process activity. Each facility waste is placed in the table by category of "Wastestream" (Column 1) and "Process Steps" (Columns 2-5). Ancillary operations (Column 6) show examples of wastes from steam generation, cleaning, and maintenance.

3.3.2   Completing a Waste Analysis For Your Facility

This section provides a blank process waste analysis process worksheet (**Figure 3-3**) and a blank waste analysis table (**Table 3-3**) for you to complete for your facility. These tools can help you identify your facility's inputs and corresponding regulated outputs (e.g., wastes), as well as the environmental regulations that apply to these wastes. Please modify these tools as needed based on your facility's operations.

To complete your waste analysis, follow the steps below. Fill the information in **Figure 3-3** and **Table 3-3** as you go through each step:

(1) Identify Process Activities and Ancillary Operations. Identify all of your process activities and ancillary operations.

(2) Identify Inputs. Identify all of the inputs to these activities and operations. Remember that some of these inputs may be regulated under specific EPA statutes (e.g., EPCRA, CERCLA, OPA, CWA), as well as under the Occupational Safety and Health Administration (OSHA). Refer to the appropriate statute-specific section for more information. Also, for a very brief overview of spill or chemical release requirements across EPA statutes, please refer to Section 9.0. *How Do I Comply With Spill or Chemical Release Requirements?*

(3) Identify Outputs and Wastestreams. Identify your outputs and the wastestream to which each output belongs. While your facility has several types of outputs, this activity is focused on identifying those outputs which are regulated - primarily wastes. As discussed earlier, waste outputs are regulated differently depending on the wastestream (wastewater, hazardous wastes, air emissions or solid wastes) to which they belong. Once you identify the wastestream, you can refer to the appropriate EPA statute-specific section of this guide for more information on how to manage those wastes and comply with the regulations.

Table 3-2. Waste Analysis for SIC Code 203 Facility

Waste-streams	Process Activities					
	Storage	Receiving and Preparation	Processing and Filling	Packaging and Printing	Storage and Distribution	Ancillary Operations
Wastewater	• Hydraulic lift oil • Waste food residue • Outdoor maintenance spills • Vegetable oils and animal fats	• Pollutants (e.g., BOD, COD, and pH) • Caustic solution • Suspended solids • Waste product residue • Oil and grease • Spilled product	• pH • Cooking oils • Oil and grease • Cooling water (from container cooling)	• Ink and coating solvents (glycol ethers, MEK) • Metal pigment compounds • Rinse water • Used fixer • Water-based inks		• Cleaning wastewater • Pollutants (e.g., BOD, COD and TSS) • pH • Residual pesticides
Hazardous Wastes	• Vehicle maintenance waste		• Spent or unusable chemical preservative	• Solvents • Liquid flux for soldering • Waste ink • Solvent-laden rags • Used fountain solution	• Vehicle maintenance waste	• Spent solvent-based cleaning materials • Spent lab chemicals • Used oil
Air Emissions	• VOCs from waste product • Odors	• VOCs • Particulates • Odors • Chlorinated organics	• CFC, ammonia emissions from coolants and cooling processes • SOx from preserving • VOCs, odors, particulate emissions	• VOCs		• Vented lab chemicals • Fugitive emissions from cleaning materials • NOx, SO$_2$, and particulates

Figure 3-3. Process Waste Analysis Worksheet

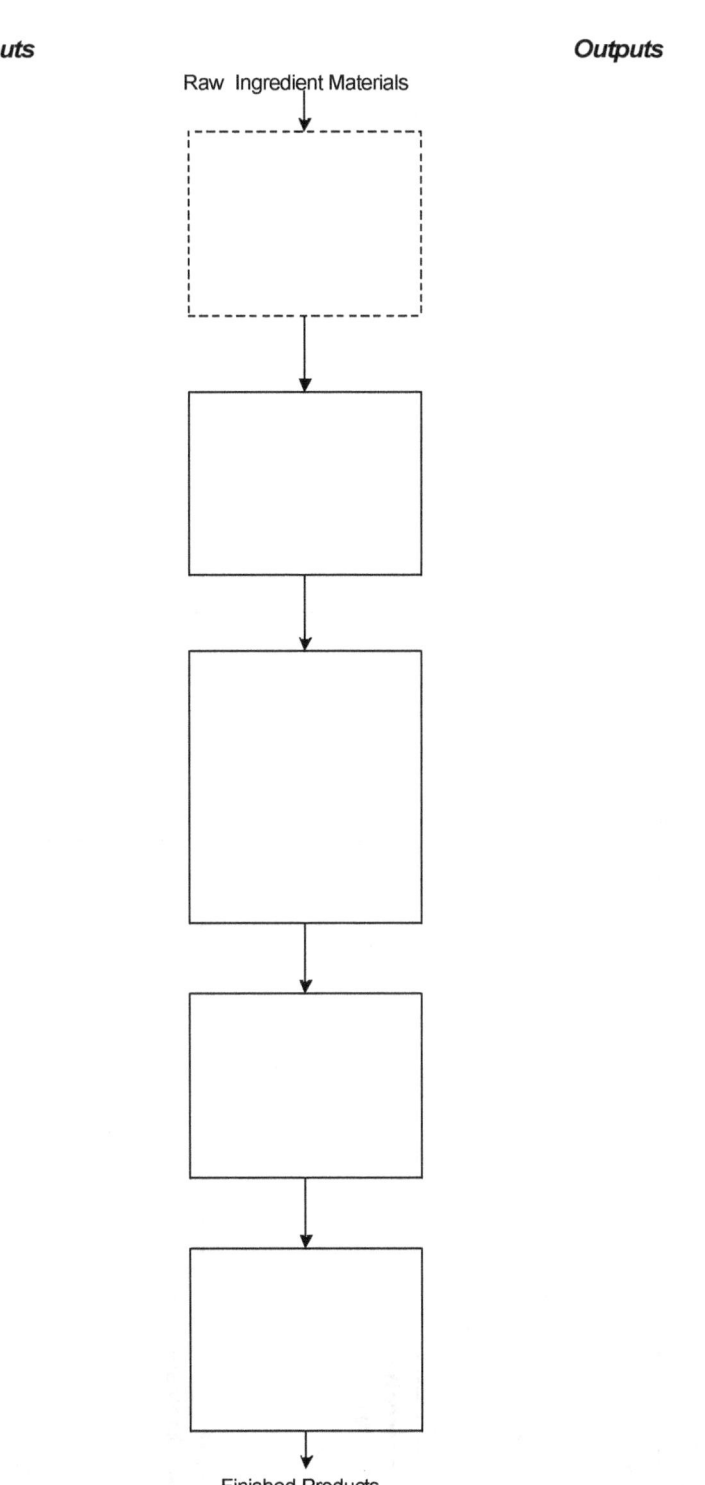

Table 3-3. Waste Analysis Worksheet

Wastestreams	Process Activities					Ancillary Operations
Wastewater						
Hazardous Wastes						
Air Emissions						
Solid Wastes						

SECTION 4 CONTENTS

4. HOW DO I COMPLY WITH WASTEWATER DISCHARGE AND RELATED REGULATIONS?

4.1 Introduction

The discharge of wastewater from your food processing facility generally will be covered by either the federal Clean Water Act (CWA) or the Safe Drinking Water Act (SDWA) (see Section 5.5 *Underground Injection Control (UIC) Requirements*. In 1972, Congress passed the Federal Water Pollution Control Act (FWPCA), now known as the CWA, which established the basic framework for protecting the waters of the United States. The CWA and its regulations now focus on keeping conventional, nonconventional (including oil and grease), and toxic water pollutants out of our rivers, lakes, and oceans.

Generally, federal regulations target three types of industrial discharges. As a food processing facility, your industrial wastewater discharges probably fall into one of these categories:

(1) **Direct discharges** which include any wastewater from an industrial facility (e.g., untreated, unpolluted wastewater or treated process wastewater) that is discharged straight to surface waters (e.g., ponds, lakes, oceans, streams, and wetlands). Storm water discharges also are considered a type of direct discharge. See Section 4.3 *Am I A Direct Discharger?*

> *If your food processing facility also operates an **Animal feeding operation (AFO)**, you may be required to obtain a National Pollutant Discharge Elimination System (NPDES) permit that includes the AFO if the AFO discharge goes directly to surface waters. Check with your permitting authority for more information.*

(2) **Indirect discharges** which include any wastewater from an industrial facility that is discharged to a publicly owned treatment works (POTW), which subsequently discharges to a surface water. See Section 4.4 *Am I An Indirect Discharger?*

(3) **Land application** of industrial wastewater discharges. Land application discharges include any wastewater from an industrial facility that is discharged to land to either condition the soil or to fertilize crops or other vegetation grown in the soil. (See Section 4.7.1 *Land Application of Wastewater*.)

For more information, visit EPA's water programs homepage at http://www.epa.gov/wow.

4.2 Wastewater Generated During Food Processing Operations

Process Wastewater

Wastewater at food processing facilities commonly is generated during food preparation, processing, and cleaning operations. As presented in Table 3-2. *Waste Analysis for SIC Code 203 Facility*, there are many common wastes that typically are found in food processing wastewater. This wastewater can contain a variety of pollutants or characteristics, some of which are regulated by federal, state, or local requirements.

Regulated pollutants include:

- biochemical oxygen demand (BOD)
- chemical oxygen demand (COD)
- total suspended solids (TSS)
- oil and grease (O&G)

- total Kjeldahl nitrogen (TKN)
- high or low pH
- ammonia nitrogen
- phosphorus.

Note that these are *examples* of conventional pollutants likely to regulated in your wastewater discharge permit. In addition, your permit may include discharges of toxics (e.g., ammonia) and nutrients (total nitrogen, total phosphorus). The amount of these pollutants that you are allowed to discharge in your wastewater will vary depending on where you discharge (e.g., direct, indirect, or other) and the applicable regulations.

Storm Water

Another potential source of wastewater at your food processing facility is storm water. Storm water discharges begin when rain comes in contact with potential pollutants, such as product spills, uncovered waste containers, or spilled liquids related to vehicle or mechanical parts maintenance. The pollutants found in storm water will be dependent on the type of material(s) the rain comes in contact with prior to discharge.

After you identify the wastewater (process and storm water) generated by your facility, you must determine how best to manage it. As discussed above, there are several methods that your food processing facility can use to dispose of your wastewater. Some of these methods require you to obtain a permit as well as conduct monitoring of pollutant levels in your wastewater. The following sections discuss the federal regulations that apply to your wastewater discharges and related activities and how you can comply with these regulatory requirements. See Section 4.3.2 *Storm Water Dischargers* for more information.

4.3 Am I a Direct Discharger?

If your food processing facility discharges process wastewater, cooling water (contact or non-contact) and/or storm water straight to surface waters (or through any conveyance system through which water flows and then discharges directly to surface waters, i.e., through a "point source"), you are a direct discharger. See Appendix A.1 *Summary of Principal Regulations Under the Clean Water Act* for the complete definition of a point source. Specific requirements that apply to food processing wastewater discharges and cooling water are discussed in Section 4.3.1; storm water discharges are discussed in Section 4.3.2.

4.3.1 Direct Dischargers to Surface Waters

As a direct discharger, you must apply for and obtain a permit under EPA's National Pollutant Discharge Elimination System (NPDES) program. A NPDES permit sets limits, often referred to as **effluent limits**, on the amounts of pollutants that can be discharged to surface waters.

Permits must be obtained from EPA or the authorized state or territory. As of March 1998, EPA has authorized 42 states and one territory to administer the NPDES program. Where permit authority has not been delegated to the state or territory, you must apply for permits directly from EPA rather than the state authority. EPA has not delegated authority to the following states and territories: Alaska, Arizona, District of Columbia, Idaho, Maine, Massachusetts, New Hampshire, New Mexico, Pacific Territories, Puerto Rico, Texas, and the federal Tribal Lands.

A NPDES permit:

- Specifies the amount of pollutants (e.g., effluent limits) that can be discharged based on either available wastewater treatment technology or on the specific water quality standards of the surface water.

> *For facilities in coastal areas, states may include stricter permit limits in order to meet the requirements of the Coastal Zone Management Act (CZMA). See Section 4.7.2 Construction or Plant Modification Activities.*

- Generally requires a facility to routinely conduct monitoring and submit reports (generally on an annual, quarterly, or monthly schedule). Such requirements are determined on a facility-specific basis; however, there are some reporting requirements that apply to all facilities. These requirements are presented in Table 4-1.

- Requires that all records related to monitoring be maintained by the facility for at least three years.

- May contain other site-specific requirements, such as (1) construction schedules, (2) best management practices (BMPs), (3) additional monitoring for non-regulated pollutants, and (4) spill prevention plans.

A NPDES permit application may be submitted as either a **general** permit or an **individual** permit, depending on EPA or state requirements. General permits, which usually are limited to storm water discharges (see Section 4.3.2 *Storm Water Dischargers*), typically are less complicated than individual permits and do not require as much information to apply for the permit. The application for a general permit is often referred to as a Notice of Intent (NOI).

How to Comply If You Are a Direct Discharger

- Contact your EPA or state regulatory agency to find out how to obtain a permit application. Apply for and obtain a NPDES permit.

- As part of the permit application, you will be required to analyze your industrial wastewater for BOD, COD, total organic carbon (TOC), TSS, ammonia (as N), temperature and pH.[1] In addition, your food processing facility will likely be required to analyze your industrial wastewater for oil and grease, and may be required to analyze for additional parameters (e.g., total phosphorus or total nitrogen) based on the water quality standards applicable to the receiving water, and any applicable state regulations. While the effluent limits and other requirements in your permit will be specific to your facility, all permits will require reporting, sample collection, and sample analysis (see 40 CFR 122.41, 136.1-136.4, and 136.3).

- • Read all permits carefully and make checklists of requirements.

- Follow the monitoring and reporting activities specified in your permit. Compare the monitoring results to the effluent limits to verify that your facility meets the effluent limits in your NPDES permit. Conduct any additional required reporting and recordkeeping activities for your wastewater discharge.

- Notify the permitting authority as indicated in Table 4-1 *Reporting Requirements for All NPDES Permit Holders*.

Table 4-1. Reporting Requirements for All NPDES Permit Holders

Specific Requirement	
Notify the permitting authority of any noncompliance with your permit that may endanger health or the environment.	Within 24 hours of becoming aware of violation; written submission within 5
Notify the permitting authority of any planned physical alterations or additions to the facility.	As soon as possible.
Notify the permitting authority of any planned changes in your discharge that may result in noncompliance.	In advance of changes.
Notify the permitting authority of the transfer of the facility to a new owner.	As soon as possible in advance of the transfer.

---

[1] Some industrial sectors also are required to analyze for some or all of the 126 priority pollutants (40 CFR 423, Appendix A); however the federal NPDES regulations do not require food processors to analyze for these pollutants.

Surface waters in the United States are protected through state-established **water quality standards**. These standards were created for the purpose of establishing minimum water quality requirements for surface waters. Water quality standards contain two distinct elements: 1) use designations, and 2) specific water quality criteria to protect these designated uses. Each water body in the country is given one or more water use designations, such as aquatic life warmwater habitat, primary contact recreation, or public water supply. Based on these designations, parameter-specific criteria are applied to the point sources discharging into the specific water body.

Total Maximum Daily Loads (TMDLs) focus on restoring and protecting surface water. TMDLs impose water quality-based discharge limits from point sources based on a watershed approach. They are written, quantitative evaluations of water quality problems and contributing sources of pollution. A TMDL:

- • Identifies the amount a pollutant needs to be reduced to meet water quality standards

- • Allocates pollutant load reductions among pollutant sources in a watershed

- • Provides the basis for taking actions needed to restore/protect a waterbody.

Direct discharges of some pollutants from your facility (e.g., ammonia) may be affected by the development and implementation of TMDLs by the states. For more information on TMDLs, refer to *Final Report of the Federal Advisory Committee on the Total Maximum Daily Load (TMDL) Program* (July 28, 1998) on the EPA Hompage at http://www.epa.gov/OWOW/tmdl/advisory.html. For additional information on pollutants of concern, water quality standards, TMDLs, or your permit, contact your permitting authority.

4.3.2   Storm Water Dischargers

Introduction and Background

Under Phase I of the storm water program, which currently is being implemented, storm water discharges associated with industrial activity, such as food processing, must be covered by a NPDES storm water permit regardless of whether they discharge to a municipal separate storm sewer system or directly to waters of the United States. Municipal separate storm sewer systems are designed to convey storm water from impermeable areas to bodies of water. Waters of the United States are defined to include virtually any surface waters, whether navigable or not.

> **Exemption from Storm Water Permit**: *Food processors that have no exposure of materials and activities to storm water are exempt from these requirements. "No Exposure" means that there is **no possibility** of storm water, snow fall, snow melt, or storm water "run on" coming in contact with any process or storage related activity.*

The term "storm water discharge associated with industrial activity" means a storm water discharge from one of 11 categories of industrial activity defined in 40 CFR 122.26. Six categories are defined by Standard Industrial Classification (SIC) codes and five are defined by regulated industry activity narrative descriptions. Food processing facilities are listed in category xi. **This category includes facilities with storm water discharges from areas where material handling equipment or activities, raw materials, intermediate products, final products, waste materials, byproducts, or industrial machinery are exposed to storm water**. These areas may include:

- Industrial plant yards
- Material handling sites
- Refuse sites
- Sites used for the application or disposal of process wastewater (as defined in 40 CFR 401)
- Sites used for the storage and maintenance of material handling equipment
- Sites used for residual treatment, storage, or disposal
- Shipping and receiving areas
- Manufacturing buildings
- Storage areas (including tank farms) for raw materials, and intermediate and finished products
- Areas where industrial activity has taken place in the past and significant materials remain.

Material handling activities at your facility include the storage, loading and unloading, transportation, or conveyance of any raw material, intermediate product, finished product, by-product, or waste product. The term excludes areas located on facility property separate from the facility's industrial activities, such as office buildings and accompanying parking lots, as long as the drainage from the excluded areas **is not mixed** with storm water drained from the above described areas. If storm water from your food processing facility is discharged to a municipal combined sewer system, the storm water discharges are subject to indirect discharger requirements (see Section 4.4).

Storm Water Permits

Food processing facilities with storm water discharges must be covered by a NPDES permit regardless of whether they discharge to a municipal separate storm sewer system or directly to waters of the United States. **Storm water permits are not required where runoff flows through a combined sewer to a POTW.** Storm water permits provide a mechanism for monitoring the discharge of pollutants from these sources to waters of the United States and for establishing appropriate controls.

Facilities can comply with NPDES permit requirements for storm water discharges by submitting (1) a **Notice of Intent (NOI)** to be covered under a **general** permit (Baseline or Multi-Sector); or (2) an application for an **individual** permit; and (3) complying with all of the conditions specified in the applicable permit. In the past, facilities could submit an application to be covered under a group permit, but this option and the original group permit have expired.

- ***General Storm Water Permit Applications - Baseline or Multi-Sector.*** Your food processing facility may be permitted under a general permit, **whether EPA or the State is the permitting authority**. General permits require development of a storm water pollution prevention plan (SWPPP) and periodic discharge monitoring. In those states and territories without NPDES authorization (i.e., EPA is the permitting authority, see below), EPA has developed and finalized general permits that cover 29 industry categories including food processors.

EPA has developed two different general permits under which food processors can discharge their storm water: the **Baseline** general permit, and the **Multi-Sector** general permit. The Baseline general permit was originally issued in 1992 and covered storm water discharges from many different facilities, both industrial and municipal, under the same requirements. The Multi-Sector permit was issued in 1995, and like the Baseline permit, covered many different industrial facilities. The main difference between the two permits is that the Multi-Sector permit established different requirements for different industries, while the Baseline established one set of requirements for all industries. As of September 1998, most food processing facilities were covered under the Multi-Sector permit. More detail on the eligibility, deadlines, expiration dates, and permit requirements of each of these facilities is provided below.

•• *Individual Storm Water Permit Applications.* If a facility has storm water discharges and did not participate in a group application, or did not obtain coverage under a general permit by March 1996, it may be required to obtain and submit an individual permit application consisting of Form 1 (General Information) and Form 2F (Application for Permit to Discharge Storm Water Discharges Associated with Industrial Activity). These forms can be obtained from and submitted to the permitting authority. Form 2F requires the facility to submit a site drainage map, a narrative description of the site identifying potential pollutant sources, and quantitative testing data of pollutant sources. A facility is required to submit an individual permit application 180 days before starting a new discharge.

Where do I get a storm water permit? General permits, NOIs, individual permit applications, and individual permits can be obtained from your NPDES permitting authority. Copies of general permits and NOIs can be downloaded from the Internet. Information on downloading NOI's and general permits can be found at **http://www.epa.gov/earth1r6/6en/w/sw/home.htm/**.

As of March 1998, 42 states and one territory have been delegated authority by EPA to administer the NPDES program. EPA has not delegated NPDES authority to the following states and territories: Alaska, Arizona, District of Columbia, Idaho, Maine, Massachusetts, New Hampshire, New Mexico, Pacific Territories, Puerto Rico, Texas, and the federal Tribal Lands. Of the delegated NPDES states and territories, only the Virgin Islands has not been delegated authority for the storm water general permits program as well. Where permit authority has not been delegated to the state or territory, food processing facilities must obtain NOIs, general permits, or individual permit applications directly from EPA rather than from the state authority. Contact your permitting authority, either EPA or your state, to find out how to obtain the appropriate documents and to determine whether the individual or general permit is required.

> *Note: If your storm water discharges are currently covered by a general permit (Baseline or Multi-Sector), you are not required to submit an individual permit application (provided that neither EPA nor the implementing agency required an individual permit application on a case-by-case basis). NOI requirements for general permits usually address only general information and typically do not require sample*

Storm Water Permits - Conditions and Requirements

Your food processing facility will be subject to different requirements depending on whether you are covered under a Baseline general storm water permit, a Multi-Sector general storm water permit, or an individual storm water permit. Conditions and requirements for each of these permits are described below.

Eligibility, Application Deadlines, and Expiration Dates

Each of the permits available to food processing facilities has different eligibility requirements, application deadlines, and expiration dates. These requirements and deadlines are summarized in Table 4-2. *Eligibility, Deadlines, and Expiration of General and Individual Permits for Food Processing Facilities.*

Baseline General Permit: In order to be covered under a storm water **Baseline general permit** on September 30, 1998, your food processing facility was required to do the following: 1) submit a NOI to EPA or the permitting authority **prior to October 1, 1992, and** 2) submit a second NOI **prior to September 9, 1997**, in order to be covered under the **administratively-extended** Baseline general permit. If you were eligible and met these deadlines, then your facility is still covered under the administratively-extended Baseline general permit.

- Eligibility: Your facility was eligible to submit these NOIs if it did **not have an adverse impact on endangered species**, and your storm water discharges were not subject to EPA's *Storm Water Effluent Limitation Guidelines*.

- Expiration: Your Baseline general permit **expired in 1997,** unless you met the deadline for submitting the second NOI to obtain coverage under the administratively-extended Baseline permit. The administratively-extended permit **will expire 90 days** after the modified Multi-Sector General permit becomes final (late 1998 or early 1999).

- • Future Requirement for Permit Coverage: The **administratively-extended Baseline general permit** expires **90 days after EPA issues modified Multi-Sector permit becomes final** (late 1998 or early 1999). When the administratively-extended Baseline expires, food processing facilities that are covered under the administratively-extended Baseline permit will have 90 days to submit a NOI to be covered under the modified Multi-Sector permit and to **meet all of the conditions** of the Multi-Sector permit (including implementing new components of the SWPPP).

EPA's recommendation to food processing facilities: Because the modified Multi-Sector permit is more complex than the Baseline permit, and because of the short time period available to implement the Multi-Sector requirements (see below), **EPA strongly recommends that food processing facilities that are covered under the Baseline permit begin to implement the requirements of the modified Multi-Sector permit (which, for food processing facilities, will have the same requirements as the currently available Multi-Sector permit) as soon as possible, rather than waiting for the Baseline permit to expire**. (See following subsection for explanation and requirements pertaining to eligibility and applications deadlines under the **Multi-Sector** and the **modified Multi-Sector general permit**.)

Table 4-2. Eligibility, Deadlines, and Expiration
of General and Individual Permits for Food Processing Facilities

	Baseline General Permit	Multi-Sector General Permit	Individual Permit
Eligibility	Food processing facilities which do not have an adverse impact on endangered species, and do not include additional non-food processing activities (e.g., can making) that are subject to effluent guidelines.	Food processing facilities which do not have an adverse impact on endangered species and **certify there will not be an impact.**	All facilities are eligible for individual permits at the discretion of the permitting authority.
Application Deadline	Existing dischargers must have submitted a NOI by 10/1/92 and a second NOI for the administratively-extended Baseline permit prior to 9/9/1997. New dischargers must submit NOI two days prior to commencing industrial activity.	Existing permit holders and group applicants must have submitted a NOI by 12/28/96 or within 90 days of finalization of modified Multi-Sector permit. New dischargers must submit NOI two days prior to commencing industrial activity.	Existing permit holders must submit applications at least 180 days prior to the expiration of existing permit. New discharges must apply for and receive permit prior to commencement of industrial activity.
Expiration	Baseline permit expired in 1997 except for those facilities that submitted NOIs prior to 9/9/97 to continue coverage (until 90 days after the modified Multi-Sector permit is issued) under the administratively-extended Baseline permit.	Multi-sector permit expires in 2000 (even for those facilities that waited to file a NOI until modifications in the Multi-Sector permit were complete). EPA will issue new Multi-Sector permits or extend existing permit when existing permit expires.	Permits are valid for dates specified on the individual permit.

Multi-Sector General Permit: In order to be covered under a storm water **Multi-Sector general permit** on September 30, 1998, your food processing facility was required to submit a NOI to EPA and to meet certain eligibility criteria. Deadlines for submitting a NOI differed for facilities in operation prior to September 29, 1995, and those commenced operation after September 29, 1995.

Facilities in the first category (i.e., those facilities **in operation prior to September 29, 1995**) had **two** periods for submitting a NOI depending on the status of your permit coverage at the time.

Period 1: The NOI submission period was between **September 29, 1995 to December 28, 1995**, if your facility was **not** covered under the **Baseline general permit**, and met the eligibility criterion below.

<u>Period 2</u>: This NOI submission period was between **June 11, 1997 and September 9, 1997**, if your facility was covered under **the Baseline general permit**, and you wanted to obtain coverage under the Multi-Sector General permit **before** the **expiration** of the **Baseline general permit** in 1997.

Facilities in the second category (i.e., those facilities that **commenced operation after September 29, 1995**) are required to submit a NOI **at least 2 days prior** to commencing operations.

- •• <u>Eligibility</u>: In any of these situations, your facility was required to meet the applicable deadline and meet both of the following condition: 1) to have no adverse impact on endangered species; **and** 2) to certify that there will be no adverse impact.

- •• <u>Expiration</u>: The Multi-Sector permit expires on September 29, 2000.

- • <u>Future Requirement for Permit Coverage</u>: If your facility is currently, or soon will be covered under the Multi-Sector permit, you must contact your permitting authority at least 90 days prior to September 29, 2000 in order to determine if you will be required to submit a NOI to be covered under a new Multi-Sector permit, or to be covered under an extension of your current Multi-Sector permit.

Individual Storm Water Permit: In order to be covered under an **Individual storm water permit,** your facility is required submit a permit application to the EPA or state permitting authority, receive a permit, and meet the conditions of the permit prior to commencing operations.

- •• <u>Eligibility</u>: Facilities are eligible for individual storm water permits if they choose to apply for individual permits, or if required to do so by the permitting authority. To be eligible, facilities that commenced operations prior to October 1992 were required to submit a storm water permit application prior to October 1992. Facilities that commenced operations after October 1992 are required to submit a permit application and receive a permit prior to commencing operations.

- • <u>Expiration</u>: Individual storm water permits expire on the date indicated on the permit. Individual permits are usually valid for five years.

- • <u>Future Requirement for Permit Coverage</u>: If your facility is currently covered under an Individual storm water permit, you are required to submit an application to renew your permit at least 180 days prior to the expiration of your current permit.

What do I do if my food processing facility missed the previous deadlines for submitting NOIs? If your facility did not submit NOIs prior to the applicable deadlines, you must contact your State or EPA Regional permitting authority to determine if you are eligible for a Multi-Sector or individual storm water permit.

What do I do if my food processing facility was previously covered under a Group permit? If your food processing facility was previously covered under a group permit, you were required to submit a NOI for the Multi-Sector permit prior to March 29, 1996 and meet the storm water pollution prevention plan conditions (see Table 4-3. *SWPPP Requirements for General and Individual Permits for Food Processing Facilities* and accompanying text below) of the Multi-Sector permit by September 25, 1996.

Table 4-3. SWPPP Requirements for General and Individual Permits for Food Processing Facilities

Baseline General Permit	Multi-Sector General Permit	Individual Permit
Requires consideration of *generic* pollution prevention measures and BMPs. Facilities subject to EPCRA Section 313 reporting requirements are required to incorporate additional measures into SWPPP, and have the plan certified by a *Professional Engineer* every three years. Does not provide guidance on specific Food Processing industry BMPs.	Requires consideration of *generic* BMPs <u>and</u> practices *specific* to the Food Processing Industry. Facilities subject to EPCRA Section 313 reporting requirements are required to incorporate additional measures into SWPPP, however, the plan only needs to be certified by the *facility operator*. Fact sheet describes applicable BMPs for the Food Processing industry.	SWPPP requirements are included at the discretion of the permitting authority.

Storm Water Pollution Prevention Plan (SWPPP). If your food processing facility is required to obtain a storm water permit, you will likely be required to prepare and implement a storm water pollution prevention plan. Facilities are required to develop SWPPPs to prevent storm water from coming in contact with potential contaminants. Each plan is facility-specific because every facility is unique in its source, type and volume of contaminated storm water discharges. Regardless of the variations, all plans must include several common elements, such as a map and site specific considerations. The elements include:

- Facility size and location

- A description of the volume of storm water and pollutants that could potentially be discharged.

- Hydrogeology

- The environmental setting of each facility

- The predicted flow of storm water discharges.

- Climate.

Storm water pollution prevention (P2) plans also must address how your facility will complete the following activities:

- Develop general and specific measures and controls to prevent or minimize pollution of storm water (articulated as BMPs in your plan)

- Develop a P2 Team

- Train employees

- Conduct inspections and evaluations

- Test outfalls

- Conduct recordkeeping.

Some SWPPP elements may be specific to the type of permit at your facility. Table 4-3 compares SWPPPs of Baseline general, Multi-Sector general, and Individual storm water permits.

Best Management Practices. BMPs are measures and controls used to prevent or minimize pollution. The most effective BMPs for reducing pollutants in the storm water discharges from your food processing facility are exposure minimization (preventing/minimizing storm water contact with potential contaminants) and good housekeeping. Exposure minimization practices reduce the potential for storm water to come in contact with pollutants. Good housekeeping practices ensure that the facility is responsive to routine and non-routine activities that may decrease exposure of storm water to pollutants. One simple practice is to move product storage, loading, or waste areas to existing enclosed structures.

While exposure minimization usually can be accomplished by good housekeeping and covering or bringing potential pollutants inside a facility, some food processing facilities may be required to develop additional structural controls to prevent contaminants from reaching storm sewers. Such controls may include cement pads, berms/dikes, screens, or separators. In a few instances, more intensive BMPs (e.g., detention ponds, filtering devices) may be necessary at your facility depending on the type of discharge, types and concentrations of contaminants, and volume of flow. Many food processing facilities already have some of these controls in place as part of spill control plans required under the Oil Pollution Act (see Section 4.6 *How Do I Comply With Oil Pollution Prevention Regulations?*), and can use these existing controls to help them meet the BMP requirements of their SWPPPs.

As part of the Multi-Sector general storm water permit, EPA has identified BMPs specific to the food processing sector. General storm water BMPs practices specific to food processors are shown in Table 4-4. *General Storm Water BMPs Required for Permit Holders in SIC Code 20.* Some or all of these BMPs may be required by your permit for your food processing facility. Even if not required, implementing these BMPs at your facility can be a low cost way to reduce contaminants in your storm water discharges.

Table 4-4. General Storm Water BMPs Required for Permit
Holders in SIC Code 20

Activity	Management Practice
Raw material unloading/product loading	• Reduce or repair defective containers (bags, drums, bottles, crates). • Prevent spills and leaks (tanks and rail cars) from reaching storm drains. • Ensure connections (hose and coupling) are secure and not leaking prior to loading or unloading. • Reduce washwater usage and cover storm drains during wash down of loading/unloading area.
Tank storage of liquids	• Conduct preventive maintenance and inspections of tanks and piping to avoid external corrosion and structural failure. • Train operators and monitor loading/unloading to prevent spills and overflows due to operator failure. • Whenever possible, provide secondary containment tanks and storage areas.
Drum and container storage of liquids	• Keep containers closed except when loading or unloading. • Keep containers covered to avoid exposure to storm water. • Regularly inspect containers for signs of corrosion, spills, or leaks. • Train operators on proper handling and transporting techniques. • Immediately clean any known spills or leaks.
Solids storage in silos, holding bins, fiber drums, etc.	• Inspect and maintain good housekeeping practices to avoid storm water contamination from spilled product, dust, or particulates.
Air emissions	• Inspect emissions from ovens or other vents that may emit solids. • Regularly clean any roof top solids, or oils and greases from air emissions. • Practice proper handling techniques to avoid dust or fine solids emissions.
Solid wastes	• Regularly inspect dumpsters and trash cans, spent equipment, scraps, etc. • Keep all wastes covered from storm water.
Wastewater	• Monitor treatment processes to prevent hydraulic overflow. • Inspect and conduct preventative maintenance on outside piping and connections.
Pest control	• Follow proper application techniques in outside areas.
Improper connections to the storm sewer	• Conduct a comprehensive site inspection to ensure that no process wastewater, process floor drains, sanitary sewers, or UST overflows are connected to storm drains.
General storm water management	• Implementation of traditional storm water management measures (oil/water separators, vegetative swales, and detention ponds).
Erosion prevention	• Identification of areas with high potential for erosion and stabilization measures or structural controls used to limit erosion.

Monitoring. Under your general or individual storm water permit, your facility will be required to conduct monitoring, which may include visual examination of storm water discharges and/or analytical monitoring. Monitoring is required primarily to provide your facility with a means for assessing your storm water contamination and evaluating the performance of your SWPPP. As indicated in Table 4-5. *Monitoring Requirements for All Food Processors* and Table 4-6 *Additional Monitoring Requirements for Specific Food Processing Operations*, most food processing facilities are required to conduct only visual monitoring if covered under the Multi-Sector general permit, and no monitoring if covered under the Baseline general permit. **Note that some states may require food processing facilities to monitor additional parameters.**

Table 4-5. Monitoring Requirements for All Food Processors

Category of Food Processor	Baseline Permit (one time monitoring for discharge characterization)	Multi-Sector Permit (ongoing monitoring)			Individual Permit
		Parameter	Monitoring Frequency	Years monitoring required	
All food processors	none* (except for animal handling/meat packing as noted below) NOTE THAT ONGOING MONITORING MAY BE REQUIRED ON A CASE BY CASE BASIS.	Visual examination	Quarterly	All	Monitoring Requirements established by the individual permitting authorities.
Facilities subject to EPCRA Section 313 reporting requirements for water priority chemicals.	O&G, BOD, COD, TSS, TKN, TP, pH, acute whole effluent toxicity, and any Section 313 water priority chemical for which the facility reports.	No additional monitoring based on EPCRA reporting status.			All individual permit requirements are specific to the facility and established by the permitting authority.
Facilities that can certify there are no significant materials or industrial activities exposed to storm water.	Exempt from all monitoring.	No provisions for these facilities in Multi-Sector permit.			
Facilities that can certify that there are no sources of a pollutant present.	No provisions for these facilities in Baseline permit.	Facilities may be exempt from monitoring certain pollutants on a pollutant by pollutant basis.			

Table 4-6 Additional Monitoring Requirements for Specific
Food Processing Operations

Category of Food Processor	Baseline Permit (one time monitoring for discharge characterization)	Multi-Sector Permit (ongoing monitoring)			Individual Permit
		Parameter	Monitoring Frequency	Years monitoring required	
Grain mill products	None	TSS	Quarterly	Year 2 and 4 of permit coverage	All monitoring required under individual permit is facility specific and determined by the permitting authority.
Fats and oils	None	TSS, BOD, COD, and nitrate + nitrite nitrogen	Quarterly	Year 2 and 4 of permit coverage*	
Animal handling/meat packing	grab and composite (where appropriate) for BOD, oil and grease, COD, TSS, TKN, Total Phosphorus, pH, and fecal coliform.	No specific requirements under Multi-Sector permit.			

* Facilities with pollutant concentrations lower than benchmarks in the second year are exempt from monitoring in the fourth year.

4.4 Am I an Indirect Discharger?

If you are an indirect discharger, your food processing facility discharges wastewater into a sewer system that leads to a municipal treatment plant, also known as a POTW. The POTW typically is owned by the local municipality or a regional board or sewer authority.

Usually, POTWs treat domestic household wastes using biological treatment processes. Most POTWs cannot handle large quantities of industrial wastewater, because certain pollutants present in industrial discharges can adversely affect the POTW's treatment processes or pass through the plant directly to surface water without receiving adequate treatment. However, some POTWs are designed to accept large industrial waste loadings, often with prearranged industrial financial assurances.

In response to the potential problems caused by industrial wastewater, federal pretreatment regulations were developed to prevent the discharge of pollutants to the POTW that will:

- Interfere with the operation of the POTW
- Pass though the POTW untreated
- Create problems with disposal of sludge from the POTW
- Cause problems to treatment plant workers from exposure to chemicals.

These regulations, referred to as the **pretreatment regulations**, apply to all industrial facilities, including food processing facilities, that discharge industrial wastewater to POTWs. Local POTWs with approved pretreatment programs have responsibility for enforcing pretreatment requirements. Because the primary enforcement authority for pretreatment regulations is often the local POTW, **to assure compliance you must contact your local POTW**, even if you have already contacted the State or EPA region.

4.4.1   Pretreatment Requirements

There are three types of pretreatment requirements: requirements for general industry (*general pretreatment standards*), requirements for specific industries (*categorical pretreatment standards*), and locally established requirements for specific facilities (*local limits*). At a minimum, the federal *general pretreatment standards* apply to your food processing facility's discharge to a POTW, while *local limits* may also apply to your facility.

(1)   **General pretreatment standards** establish minimum discharge requirements for all industrial discharges, including those from food processing facilities. These standards protect POTWs by prohibiting specific wastestreams from being discharged by industrial users. The general types of pollutants prohibited under the general pretreatment standards include:

  •   Pollutants that cause pass through or interference at the POTW (including vegetable oils and animal fats).

  ••   Pollutants creating a fire or explosion hazard in the POTW.

  •   Pollutants that will cause corrosive structural damage (i.e, any wastewater with a pH less than 5).

  ••   Pollutants that are solid or viscous that can obstruct the wastewater flow.

  •   Any pollutant released in a discharge at a flow rate or concentration that will cause interference at the POTW.

  •   Heat in amounts that will inhibit biological activity at the POTW, but in no case, discharges that will cause the POTW influent to exceed 104 degrees Fahrenheit.

  •   Petroleum oil, non-biodegradable cutting oil, or products of mineral oil in amounts that will cause interference or pass through.

  •   Pollutants that result in the presence of toxic gases, vapors, or fumes in the POTW that may cause acute worker health and safety problems.

  •   Any trucked or hauled pollutants, except at discharge points designated by the POTW.

Some examples of typical pollutants specific to SIC Code 20 that are prohibited under the general pretreatment standards include pollutants discharged at a flow rate or concentration that will cause interference at the POTW (i.e., high BOD and COD loads). In cases where there are high BOD or COD loads in the wastewater, the POTW may choose to include these discharges in their surcharge program (see Section 4.4.2).

(2) **Categorical pretreatment standards** are standards established for specific types or categories of industrial facilities or processes. These facilities are known as categorical industrial users. EPA does not consider food processing facilities to be categorical industrial users. EPA has not established specific numerical limits for indirect discharges from food processors. Hence, **categorical pretreatment standards** that apply to food processing operations require compliance with 40 CFR 403 (the general pretreatment standards).[2]

(3) **Local Limits or Requirements** are standards established by the POTW for any or all of the industrial facilities from which it receives wastewater. These limits are designed to protect the POTW and its workers, and to meet the POTW's own direct discharge permit limits. POTWs often require food processing facilities to clean and maintain grease and grit traps on a specified schedule.

> **Can making operations.** *If your food processing facility also includes can making (not simply packaging materials in cans), your facility will be subject to additional categorical pretreatment standards. These standards can be found in 40 CFR 465.40 through 465.46. In addition, your facility will be considered a Significant Industrial User (SIU) and be subject to the permit criteria discussed below.*

Where the POTW local limits are **more** stringent than federal requirements, they will **replace** the federal requirements. As a food processing facilty, your POTW may or may not have local limits. For specific POTW limits that apply to your facility, you must contact your local POTW. **Even if your facility is not subject to local limits, the general pretreatment standards do apply.**

---

[2] EPA does not consider food processors involved in dairy products processing, grain mills, canned and preserved fruits and vegetables, canned and preserved seafood, sugar processing, and meat products point source categories (see 40 CFR 405-409 and 432) to be categorical industrial users. EPA has not established specific numerical limits for indirect discharges, known as categorical pretreatment standards, for these facilities. (Memo from Director Office of Water Enforcement and Permits to Regional Water Management Divisions Directors and Regional Water Compliance Branch Chiefs, 2/16/89)

40 CFR 403 provides **General Pretreatment Standards** for various types of food processing facilities. In 40 CFR 405-409 and 432, EPA provides several tables of effluent limits for various types of discharges (e.g., direct and indirect discharges from new and existing facilities) from various types of food processing facilities. There is a different table for each type of food processing facility, and each type of discharge from that type of facility; the tables simply refer to the **General Pretreatment Regulations**, or indicate that the industrial users (food processors) **must comply with 40 CFR 403 (the general pretreatment standards).** If EPA considered food processing facilities to be categorical industrial users, then EPA would have established specific numerical limits for indirect discharges from food processors. Thus, the tables simply provide a reminder that food processors must comply with General Pretreatment Standards, rather than impose additional categorical pretreatment requirements.

Permit Criteria

In addition to the local limits, your POTW may require you to have a wastewater discharge permit that requires you to monitor, submit reports, and keep records of your industrial wastewater discharges. POTWs are required by federal law to permit significant industrial users (SIUs), and may permit any or **all** of their industrial dischargers, not just significant industrial users as defined below. Your food processing facility is considered a SIU if it meets any of the following criteria:

- Is subject to categorical pretreatment standards (not applicable to SIC 20);

- Discharges an average of 25,000 gallons or more per day of process wastewater (excluding sanitary, noncontact cooling, and boiler blowdown wastewater) (may apply to a food processing facility);

- Contributes a process wastestream which makes up five percent or more of the average dry weather hydraulic or organic capacity of the POTW treatment plant (may apply to a food processing facility); or

- Is designated on the basis that the industrial discharger has a reasonable potential for adversely affecting the POTW's operation or for violating any pretreatment standard or requirement (may apply to a food processing facility).

If your food processing facility meets any of the above criteria, you must contact the POTW and receive a permit prior to discharging wastewater. However, even if a permit is not required, you will need to get *approval* from the POTW for your industrial wastewater discharge to the POTW.

Additional Requirements

Usually, the monitoring, reporting, and recordkeeping requirements, as well as the wastewater discharge limits, are specified in facility-specific permits. Also, as shown in Table 4-7 *Reporting Requirements for All Indirect Dischargers*, there are some reporting requirements, that apply to **all** indirect dischargers (even if they do not have a permit).

Table 4-7. Reporting Requirements for All Indirect Dischargers[3]

Requirement	Time Frame
Notify the POTW or State of a discharge of wastewater that could cause problems to the POTW, including slug loading. A slug loading is defined as any relatively large release of a pollutant that might not ordinarily cause a problem when released in small quantities.	Immediately (40 CFR 403.12(f)).
Notify the POTW or state of substantial change in wastewater discharge.	Prior to the change.
Notify the POTW, state hazardous waste authorities and EPA Regional Waste Management Division Director of a discharge of hazardous waste. This is a one-time notification required of those who discharge more than 15 kg of a hazardous substance in a month; or if the substance is acutely hazardous and any amount is discharged. *Note: A list of acutely hazardous wastes can be found in 40 CFR 261.30(d) and 261.33(e).* The written notification must include (1) the name of the listed hazardous waste (as listed in 40 CFR 261); (2) the EPA hazardous waste number; (3) the type of discharge; and (4) a certification that a program is in place to reduce the amount and toxicity of the hazardous waste that is generated, to the degree that is economically feasible. If discharging more than 100 kg of hazardous waste in one month, the notification also must include (1) identification of the hazardous waste constituents that are contained in the waste; (2) an estimate of the mass and concentration of the constituents in the waste stream discharged during the month; and (3) an estimate of how much will be discharged in the next 12 months. *If any new substance is listed under RCRA and a facility discharges the substance, the facility must notify the authorities cited above within 90 days of the new listing.*	One-time notification for each hazardous waste discharge [40 CFR 403.12(p)].
Additional Requirements	As specified by POTW

Private Wastewater Treatment

Sometimes, a food processing facility may discharge to a sewer that is owned and/or operated by a private industrial treatment plant. The private plant is not considered a POTW and is not required to implement a pretreatment program for indirect dischargers as described above. However, the plant may have requirements of its own that apply to the discharge of wastes.

---

[3] In addition to the requirements listed for all industrial dischargers in Table 4-7, all SIUs (and most food processors will be SIUs as a result of their high strength waste stream) are required to submit once every six months a description of the nature, concentration and flow of the pollutants required to be reported by the control authority (either the POTW or the State/EPA).

How to Comply If You Are an Indirect Discharger

- Obtain a copy of the state and/or local sewer use regulations or ordinance by contacting your state and/or local POTW to determine what requirements apply to your facility.

- Contact the POTW or state to determine whether your facility must obtain a permit. Even if you are not required to obtain a permit, you may be required to obtain **approval** for your wastewater discharge.

- • Meet, at a minimum, the federal **general pretreatment standards** if you are an indirect discharger (even if your POTW does not require a permit).

- Verify whether your wastewater discharge is meeting the effluent limits in your permit (if you have one) and that your facility is not discharging any prohibited pollutants (see Section 4.4.1 *Pretreatment Requirements*) to the POTW.

- Conduct monitoring, reporting, and recordkeeping activities for your industrial wastewater discharge. Maintain records for all samples collected for monitoring activities for at least three years. These records, which should be available for review at any time, must include:

  - Date, place, method, and time of sampling and the names of the person(s) taking the samples

  - Date(s) the laboratory performed the analyses and the analytical methods used

  - Laboratory that performed the analyses

  - Results of the analyses.

- • Notify the state or POTW (see Table 4-7):

  - Immediately of a discharge of wastewater from your facility that could cause problems to the POTW, including slug loading

  - Promptly prior to any substantial change in your wastewater discharge

  - Of a hazardous waste discharge from your facility. You also are required to notify state hazardous waste authorities and the EPA Regional Waste Management Division Director.

4.4.2   Calculating Your Surcharge

Even if permits are not required, wastewater treatment by POTWs costs money, and most treatment works charge according to the volume of sewage treated. Many POTWs charge flat rates per unit flow and pollutants, regardless of concentration. Other POTWs may charge extra if the waste load exceeds certain specified levels. This extra charge is called a surcharge. Surcharges are used for pollutants that typically can be treated at the wastewater treatment plant such as BOD and TSS (two pollutants commonly found in high concentrations at food processing facilities).

By definition, a surcharge is a charge that is based on the pounds of waste material in industrial wastewater **in excess** of normal levels. The surcharge is levied in addition to the normal sewer service charge which is the regular charge for treating normal strength wastes and is generally based on volume alone. Because a surcharge typically is based on the pounds of waste above "normal," there is an economic incentive for facilities to reduce the strength of these wastes. An example of how to calculate a surcharge is provided below. Note, even if a POTW uses a surcharge system, they will also impose a limit, above which you cannot discharge.

Example Surcharge Calculation[4]. The total amount of BOD in your food processing facility's wastewater (BOD loading) can be calculated by multiplying the BOD concentration by the amount of effluent as follows:

Amount of BOD = effluent (million gallons) x BOD concentration x 8.34 (conversion factor)

The monthly surcharge is based on the amount that the BOD concentration exceeds a specified level. Assume your food processing plant discharges 4.0 million gallons of wastewater per month with a BOD concentration of 2,500 mg/l and is subject to a surcharge on BOD in excess of 250 mg/l. To find the monthly surcharge cost, multiply the *excess* amount of BOD by the surcharge rate.

Amount of BOD subject to surcharge = (2,500 mg/l -250 mg/l) x 4.0Mgal/month x 8.34 = 75,060 lbs/month

If the surcharge rate is 10¢ per lb of excess BOD, the monthly cost is:

Surcharge cost = 75,060 lbs x 10¢ /lb = $7,506.00.

In addition to the charge for excess BOD and TSS, surcharges also may be used for excessively high levels of COD and TKN.

4.5 How Do I Dispose of Industrial Sludge?

Industrial sludge is defined as sludge generated at an industrial facility during the treatment of industrial wastewater with or without combined domestic sewage. The way you choose to dispose of your industrial sludge determines how it is regulated:

•• The application of industrial sludge to land is regulated under 40 CFR 257.

---

[4] North Carolina Agricultural Cooperative Extension Service. "Bank or Drain: Cut Waste to Reduce Surcharges for Your Dairy Plant," North Carolina Pollution Prevention Pays Program. CD-26. March 1996. (JWM.) **Http://www.bae.ncsu.edu/baeprograms/extension/publicat/wqwm/cd26.html/**.

- The disposal of industrial sludge in a municipal solid waste landfill is regulated under 40 CFR 258. Although pollutant limits are not imposed under this regulation, sludge to be disposed of must be nonhazardous, as demonstrated by using the Toxicity Characteristic Leaching Procedure (TCLP) and passing a paint filter test. Contact your municipal solid waste landfill for more information on industrial sludge disposal requirements.

Note: Industrial sludge produced by wastewater treatment processes at your food processing facility is **not regulated** under EPA's biosolids rule (40 CFR 503) which allows beneficial uses of sludge generated from treatment of **domestic** wastewater and sludge generated from **municipal** wastewater treatment plants.

4.6 How Do I Comply With Oil Pollution Prevention Requirements?

4.6.1 Introduction and Background

Food processors today use oils for many purposes, including cooking oil for frying, food grade hydraulic oil for moving product, and/or diesel fuel for vehicles and as backup for boilers. Also several segments of the industry refine and/or process animal and vegetable oils for food and/or non-food uses such as soaps, inks, paint or varnish, resins and plastics, lubricants, fatty acid and other products. Non-food uses of vegetable oils are increasing with a corresponding increase in the amounts of these oils being produced and/or stored at refineries and bulk transportation facilities. Facilities that store petroleum and nonpetroleum oils (e.g., vegetable oils and animal fats) must follow Oil Pollution Prevention Regulations, also known as the Spill Prevention, Control and Countermeasures (SPCC) Regulation.

Oil discharges can have a variety of impacts on terrestrial and aquatic ecosystems, and on human drinking water resources and water treatment plants. Petroleum oil spills can also create the potential for explosion and fires, that, in turn, may lead to more equipment failures, mores spills , and may endanger people as well as wildlife. Like petroleum oils, vegetable oils and animal fats and their constituents can severely harm aquatic and terrestrial organisms and their habitats, foul shorelines, clog drinking water treatment plants, upset or disable a wastewater treatment plants, and catch fire when ignition sources are present.

EPA issued the SPCC regulation in 1973 (40 CFR 112) to address the oil spill prevention provisions contained in the Clean Water Act (CWA) of 1972.[5] The main objective of the SPCC program is to prevent oil spills from regulated aboveground and underground storage tanks from reaching navigable waters of the U.S. or adjoining shorelines (see Section 4.3 *Am I A Direct Discharger?* for a definition of navigable waters). In 1990, Congress passed the Oil Pollution Act (OPA) that amended Section 311 of the CWA to require "substantial harm" facilities to develop and implement facility response plans (FRPs) (see Section 4.6.3 *Facility Response*

---

[5] EPA' s Oil Program Center, Office of Emergency and Remedial Response (OERR), is revising the SPCC regulation and expects to finalize it in mid-1999. Some SPCC criteria and other provisions may change as a result. This regulatory revision pulls together changes that EPA has proposed to the SPCC regulation in three separate Federal Register notices -- one each in 1991, 1993 and 1997.

Plans). FRPs help facility owners/operators develop a response organization and identify the resources needed to respond to an oil spill adequately and in a timely manner. A unique aspect of these regulations is that unlike many other EPA programs, EPA may not delegate implementation of the SPCC regulation and the FRP regulation to state or tribal governments. States, however, may have their own requirements; therefore be sure to contact your state agency to find out about applicable requirements.

What is an Oil?

Oils are defined under several statutes including the CWA and its amendments under the Oil Pollution Act (OPA) of 1990. As a result, overlapping regulatory interpretations exist. For this reason, the EPA and the U.S. Coast Guard (USCG) are currently developing a nationally consistent program policy and methodology for facilities to determine whether a given substance is considered an oil under the existing CWA.

Under the CWA, the definition of oil includes oil of any kind and any form, such as petroleum and nonpetroleum oils. Generally, oils fall into the following categories: crude oil and refined petroleum products, edible animal and vegetable oil, other oils of animal or vegetable origin, and other nonpetroleum oils.

Many substances are easily recognizable as oils (e.g., gasoline, diesel, jet fuel, kerosene, and crude oil). Under the CWA definition, many other substances are considered oils which may not be easily recognizable by the industry, including mineral oil, the oils of vegetable and animal origin and other nonpetroleum oils. Therefore, your facility should work closely with the EPA and USCG (if applicable) to make determinations for the substances you store, transfer, or refine.

EPA's regulation requires facilities to prepare a plan and implement measures to prevent and control spills, regardless of the cause (e.g., human operational error, equipment failure or natural causes, such as lightning striking a tank). Facilities that fully comply with the requirements reduce the number

> **Use of Terms**: *The following terms are used throughout this section: "spill," "discharge," and "release." They are used either as they appear in the regulations or as seems most appropriate for the discussion. Please refer to the regulations for specific definitions.*

and severity of discharges, thereby reducing the high costs of environmental cleanup or the additional permitting requirements that could be imposed in the event of a discharge. Facilities that are not in compliance are at greater risk to experience an oil spill that may result in a discharge into a navigable waterway or adjoining shoreline. The cost of cleanup would not only include repairing the damage to the facility (e.g., soil removal or equipment repair), but could extend beyond the facility's boundaries to affected offsite areas. Regulatory agencies may require modifications to operations or revisions to plans.

4.6.2 SPCC Requirements

Nontransportation-Related Facilities

The SPCC requirements (40 CFR 112.3 through 112.7) of the Oil Pollution Prevention regulation apply to **nontransportation-related facilities** that meet these criteria:

- Could reasonably be expected to discharge oil in harmful quantities into navigable waters of the United States or upon the adjoining shorelines, AND

- Have (1) an aboveground oil storage capacity of more than 660 gallons in a single container; or (2) a total aboveground oil storage capacity of more than 1,320 gallons; or (3) a total underground storage capacity of more than 42,000 gallons.

> *Storage Capacity: Remember, the requirements apply specifically to your storage capacity, regardless of whether the tanks are completely filled.*

Many facilities have aboveground storage tanks (ASTs) with storage capacities that meet the above criteria and therefore, must comply with the SPCC requirements. In addition to these federal requirements, there are often state and local requirements for ASTs. These requirements typically incorporate standards established by organizations such as the National Fire Protection Association and the American Petroleum Institute. Construction, design, and operation requirements for ASTs are typically governed by state and local fire marshals or an environmental officer. In addition to consulting with your fire marshal, you should also check with your state regulatory agency for information on additional AST requirements.

> *According to the SPCC proposed rule (October 21, 1991; 56 FR 54612), an AST is a tank or combination of tanks (including the connecting pipes) used to contain regulated substances that breaks the natural grade of the land.*

Your AST system(s) may not be subject to EPA's SPCC regulations (40 CFR 112) if it meets the following conditions:

- Your onshore or offshore facility which, due to its location, could **not** be reasonable expected to discharge oil into or upon navigable water of the U.S. or adjoining shorelines.

- Equipment or operations of vessels or transportation-related facilities (both onshore or offshore) are subject to the authority of the Department of Transportation, with certain exceptions.[6]

---

[6] The exceptions are that certain offshore facilities along the Great Lakes, rivers, coastal wetlands, and the Gulf Coast barrier islands are subject to EPA's Oil Pollution Prevention regulations, as a result of the Department of Interior's (DOI's) redelegation of authority for these facilities to EPA in a *Memorandum of Understanding* among EPA, DOI and DOT, effective February 3, 1994.

If your facility is subject to the SPCC requirements based on the above description, EPA requires you to prepare an SPCC plan (see below) and conduct an initial screening to determine whether you are required to develop an FRP. Those facilities that could cause "substantial harm" to the environment must prepare and submit an FRP to EPA for review (see Section 4.6.3 *Facility Response Plans*).

> *SPCC-regulated facilities must also comply with other federal, state, or local laws, some of which may be more stringent.*

The SPCC Plan

The SPCC Plan is a written site-specific description detailing how a facility's operation complies with 40 CFR 112. In order to comply with 40 CFR 112, the SPCC Plan must be a carefully thought out plan, prepared in accordance with good engineering practices and which has the approval at a level with the authority to commit the necessary resources.

Regulated facilities in existence at the time the regulation went into effect, on January 10, 1974, were required to have a Plan prepared within 6 months of the effective date of the regulation and to have fully implemented the Plan within one year of the effective date of the regulation.

Newly constructed facilities must prepare an SPCC Plan within 6 months of the date they commence operations and to fully implement the Plan within one (1) year of starting operations. If you, as facility owner or operator, are unable to implement an SPCC Plan within this time frame due to circumstances beyond your control, you may request an extension, in writing, from the EPA Regional Administrator by following the procedures explained in 40 CFR 112.3(f).

While each SPCC Plan is unique, certain elements must be included in order for the SPCC Plan to comply with the provisions of 40 CFR 112. If a section does not apply to your facility, your Plan must state this. These elements include, but are not limited to, the following:

- *Professional Engineer (PE) certification:* The SPCC Plan must be reviewed and certified by a registered PE who is familiar with the facility and with 40 CFR 112. The engineer's name, registration number and state of registration must be included as part of the SPCC Plan. The engineer's seal should be affixed to the Plan as part of the certification. By certifying the Plan, the engineer is attesting that he/she has examined the facility and is familiar with the facility, its SPCC Plan and the SPCC requirements, and that the Plan has been prepared in accordance with good engineering practices. In order to satisfy the requirements of 40 CFR 112.5, all subsequent amendments must also be certified by a PE, as described above.

- *SPCC Plan Kept Onsite:* If your facility is manned at least 8 hours a day, you are required to maintain a complete copy of the SPCC Plan onsite. If the facility is manned less than 8 hours a day, the Plan must be kept at the nearest field office that is manned. The Plan must be made available for review by the EPA Regional Administrator or his/her representative during normal business hours.

- *Management Approval:* The SPCC Plan have the full approval of management at a level with the authority to commit the resources necessary to implement the Plan. The appropriate manager's signature should be included in the SPCC plan.

- ***Plan Sequence Follows 40 CFR 112:*** The SPCC Plan must include a complete discussion of your facility's conformance with all applicable SPCC requirements and shall follow the sequence of 40 CFR 112.7. To help facilities in preparing and reviewing SPCC Plans, the EPA Oil Program developed a sample Plan that you may obtain from your EPA Regional Office.

  All spill prevention practices used at your facility must be addressed with a complete and accurate description included in the Plan. It is possible that some items in 40 CFR 112 are not applicable to your facility. Every item must be addressed in the Plan, even if your facility does not have every item. For example, under facility transfer operations, the first item requires that buried piping be protectively wrapped or cathodically protected. Some facilities may not have any buried piping. For these cases, the SPCC Plan should indicate that there is no buried piping at the facility.

- ***Spill History:*** If your facility has experienced one or more spill events within 12 months prior to January 10, 1974, you should include a written description of each spill, corrective actions taken, and plans for preventing recurrence in the SPCC Plan.

- ***Spill Prediction:*** Where industry experience indicates a reasonable potential for equipment failure, the SPCC Plan should include a prediction of direction, rate of flow, and total quantity of oil that could be discharged from your facility as a result of each major type of failure. Such failures include, but are not limited to, tank failure due to overflow, rupture or leakage; pipeline failure due to rupture or corrosion; leaking flanges, gaskets, expansion joints, valves, or catch pans; spills from bulk oil loading or unloading operations; and leaks due to other causes, such as failure of wastewater or storm water treatment or disposal systems.

  Topographic maps are often useful for predicting and illustrating the direction of flow and bodies of water which might be affected by a spill.

- ***Plan review:*** The SPCC Plan must be amended whenever there is a change in facility design, construction, operation or maintenance which materially affects the facility's potential to discharge oil into or upon navigable waters or adjoining shorelines.

  In addition, the owner or operator of a regulated facility, is required to review and evaluate the SPCC Plan at least once every three (3) years from the time the facility becomes subject to the SPCC requirements. Within six (6) months after this review and evaluation, you must amend the SPCC Plan to include more effective prevention and control technology:

  - if such technology will significantly reduce the likelihood of a spill event from the facility, **and**
  - if the technology has been field-proven at the time of the review.

  In order to satisfy the requirements of 40 CFR 112.5, all such amendments must be certified by a PE, as described earlier in this section.

- ***Amendment by EPA Regional Administrator (RA):*** After review of the SPCC Plan and any other information submitted, the RA may require you to amend the SPCC Plan if (1) it does not meet the SPCC requirements or (2) such amendment is necessary to prevent or contain future discharges from the facility. The RA will consider any recommendations made by the state agency in charge of water pollution control during the review process. If the RA proposes that the SPCC Plan be amended, the RA shall provide written notification to you specifying the terms of the proposed amendment. Upon receipt of this notification, you will then have 30 days in which to respond, in writing, to the proposal, and offer any additional information, arguments or counterproposals.

The RA will then review all available information and either notify you of any amendment required or rescind the original notice. Usually, if an amendment is required, it must be made part of the SPCC Plan within 30 days after the final EPA notice and implemented as soon as possible, but not later than six (6) months after you amend the Plan (unless the RA specifies another date). You may appeal an RA's decision regarding any amendment, in writing, to EPA's Administrator within 30 days of receipt of the RA's final notice. You must also send a copy of the appeal to the RA.

- **Secondary Containment or Contingency Plans:** You are required to install appropriate containment and diversionary structures or equipment, such as dikes, berms, and retaining walls, as described in 40 CFR 112.7, to prevent discharged oil from reaching navigable water, unless it can be clearly demonstrated that installation of such structures or equipment is not practical or practicable. Impracticability pertains primarily to those cases where severe space limitations or other physical constraints may preclude installation of structures or equipment to prevent oil from reaching navigable water. Demonstrating impracticability on the basis of economic considerations is not acceptable.

  In the event that such impracticability can be demonstrated, you must provide the following plan and resources in place of containment structures or equipment:

  - A strong oil spill contingency plan following the provisions of 40 CFR 109, and

  - A written commitment of manpower, equipment and materials required to expeditiously control and remove any harmful quantities of oil discharged.

  > *Note: FRPs developed by **substantial harm** facilities may meet the above requirements for an oil spill contingency plan (see Section 4.6.3 Facility Response Plans).*

- **Spill Reporting:** First, you must report oil releases to the **National Response Center at 1-800-424-8802 or 703-412-9810 (Washington, D.C. area)** (see Section 4.6.4 *Oil Spill Notification, Response, and Recovery*).

  In addition, the owner or operator of a regulated facility must submit, in writing, certain information including the SPCC Plan to the EPA RA within 60 days, if the release meets either of the following conditions:

  - **Either** a single discharge of more than 1,000 gallons of oil;

  - **Or**, two reportable spills/discharges of oil in harmful quantities, during any 12-month period, into or upon navigable waterways, shorelines, etc.

The required information includes:

- Name of the facility;
- Name(s) of the owner or operator of the facility;
- Location of the facility;
- Date and year of initial facility operation;
- Maximum storage or handling capacity of the facility and normal daily throughput;
- Description of the facility, including maps, flow diagrams and topographical maps;
- The cause(s) of such spill(s), including a failure analysis of the system or subsystem in which the failure occurred;
- Corrective actions and/or countermeasures taken, including a complete description of equipment repairs or replacements; and
- A copy of the SPCC Plan and any other information pertinent to the Plan or the spill(s).

A complete copy of all information sent to the RA must also be sent to the state agency in charge of water pollution control activities.

- **Performance-Based SPCC Requirements:** In addition to general requirements, the SPCC rule also has performance-based requirements in 40 CFR 112.7 for drainage control; bulk storage tanks; tank car and truck loading and unloading racks; various onshore and offshore production facility operations; onshore and offshore oil drilling, production, and workover facilities; security; and training. These specific requirements are not discussed further in this guidance. Please refer to 40 CFR 112.7 for more information.

4.6.3 Facility Response Plans (FRPs)

In 1990, Congress passed the Oil Pollution Act (OPA) which amended Section 311 of the Clean Water Act to require "substantial harm" facilities to develop and implement FRPs. EPA's FRP requirements, which were published as a final rule in the Federal Register on July 1, 1994, are codified at

> *For more information on FRPs, access EPA's Oil Program webpage at http://www.epa.gov/oilspill/.*

40 CFR 112.20 and 112.21 and include Appendices A through F. Under the FRP requirements, owners and operators of facilities that could cause "substantial harm" to the environment by discharging oil into navigable water bodies or adjoining shorelines must prepare plans for responding, to the maximum extent practicable, to the worst case discharge and to a substantial threat of such a discharge of oil.

Determination of Response Plan Applicability

Owners or operators of all facilities subject to the Oil Pollution Prevention Regulation must familiarize themselves with the rule to determine whether their facility meets the "substantial harm" criteria. Facilities subject to the FRP requirements under 40 CFR 112.20 are referred to either as **substantial harm** facilities or **significant and substantial**

> *Although you don't have to submit it to EPA, your facility should maintain the Certification of the Applicability of the Substantial Harm Criteria with your SPCC plan.*

harm facilities. EPA has different roles when handling the FRPs for these 2 categories of facilities. FRPs from substantial harm facilities **are reviewed** by EPA while FRPs from significant and substantial harm facilities are **reviewed and must be approved** by EPA (see Figure 4-1 *Determination of Response Plan Applicability*). Facilities that do not fall into these two categories do not have to submit FRPs. Under 40 CFR 112.20 (e), facilities that **do not** meet the "substantial harm" criteria must document this determination by completing the "Certification of the Applicability of the Substantial Harm Criteria Checklist," provided as 40 CFR 112, Appendix C, Attachment C-II. This certification should be maintained with the facility's SPCC plan.

Figure 4-1. Determination of Response Plan Applicability

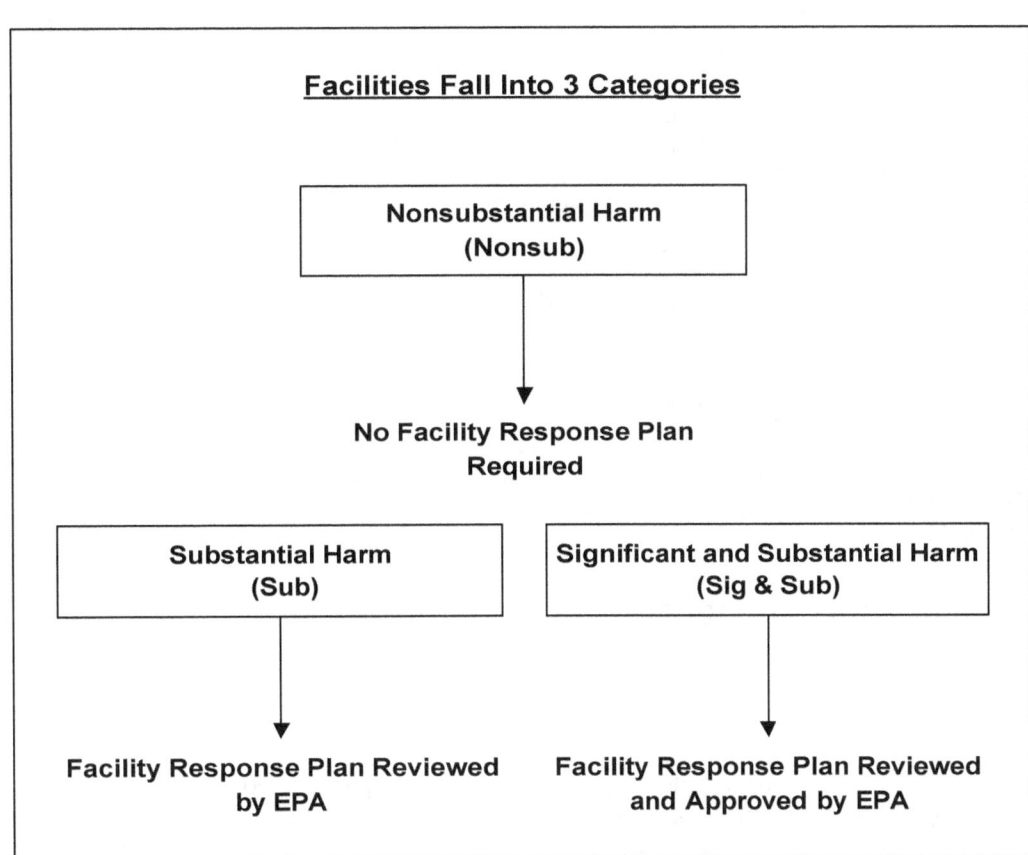

Substantial Harm Facilities

Facilities that are considered to pose a threat of substantial harm are required to prepare and submit FRPs. EPA recognizes two ways in which a facility may be identified as posing a risk of substantial harm:

- **Either** through a self-determination process (EPA has established criteria located in 40 CFR 112.20 to assist facilities in making the determination - see below),

- **Or** by a determination of the EPA Regional Administrator (RA).

As outlined in 40 CFR 112.20 (f)(1), your facility has the potential to cause substantial harm if:

•• **Either** the facility transfers oil over water to or from vessels **and** has a total oil storage capacity, including both aboveground storage tanks (ASTs) and underground storage tanks (USTs), greater than or equal to 42,000 gallons;

- **Or** the facility's total oil storage capacity, including both ASTs and USTs, is greater than or equal to one million gallons **and one of the following is true**:

  - The facility does not have secondary containment for each aboveground storage area sufficient to contain the capacity of the largest AST within each storage area plus freeboard to allow for precipitation;
  - The facility is located at a distance such that a discharge could cause injury to fish and wildlife and sensitive environments;
  - The facility is located at a distance such that a discharge would shut down a public drinking water intake; or
  - The facility has had a reportable spill greater than or equal to 10,000 gallons within the last five years.

EPA's RA may determine whether a particular facility may cause "substantial harm." EPA's RA may consider factors similar to the self-selection criteria, as well as other factors, including the type of transfer operations at a facility, the facility's oil storage capacity, lack of secondary containment, proximity to environmentally sensitive areas or drinking water intakes, and/or the facility's spill history. These factors and how they are applied are shown in Figure 4-2 *Flowchart of Criteria for Substantial Harm* shown on the next page and also in 40 CFR 112, Appendix C. The EPA RA will notify your facility if EPA has determined that your facility poses a threat of "substantial harm."

Significant and Substantial Harm Facilities

EPA is also required to identify a **subset** of substantial harm facilities that could cause **significant and substantial harm** to the environment upon a release of oil. EPA bases its determination on factors similar to the criteria used to determine substantial harm, as well as the age of tanks, proximity to navigable water, and spill frequency. Facilities are notified by EPA in writing of their status as posing significant and substantial harm. If your facility is notified by EPA, you must submit your FRP to EPA for review and approval. The RA will review your FRP and inspect your facility for viability and compliance with the regulations before EPA approves the plan.

> *Remember: FRPs from substantial harm facilities are reviewed by EPA while FRPs from significant and substantial harm facilities are reviewed and must be approved by EPA.*

Figure 4-2. Flowchart of Criteria for Substantial Harm

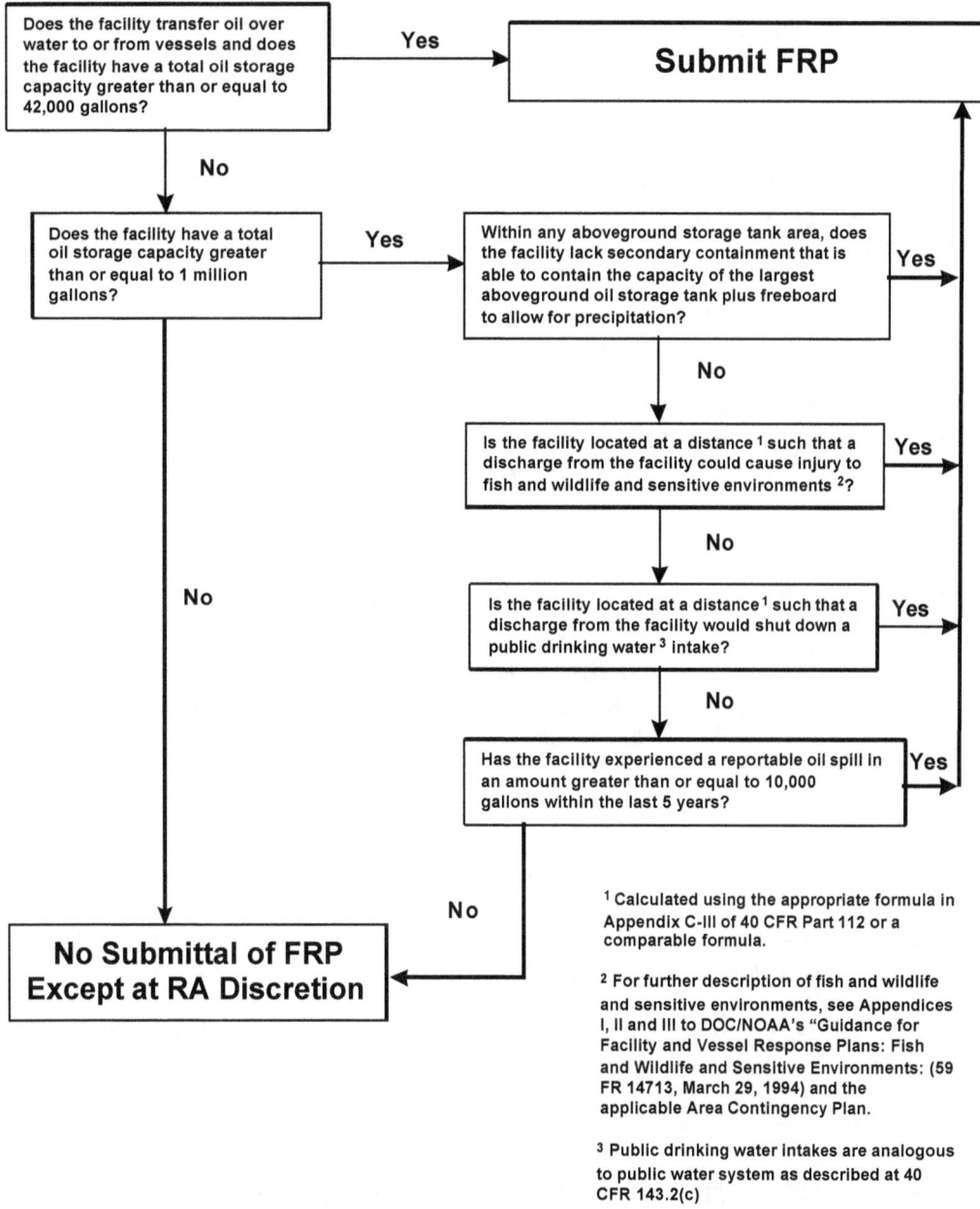

FRP Development

If it has been determined, either through the self-selection process or by notification from the EPA RA, that your facility poses a threat of "substantial harm" to the environment, you must prepare and submit a FRP to the appropriate EPA Regional Office.

To assist you in preparing a FRP, EPA has prepared and included a "model facility response plan" in 40 CFR 112.2, Appendix F. EPA recognizes that many facilities may have existing *response* plans prepared to meet other requirements. Under OPA, you do not need to prepare a separate FRP provided that your original response plan:

> *EPA also recognizes that many facilities have established SPCC plans. Although response plans and prevention plans are different, and should be maintained separately, some sections of the plans may be the same. Under OPA regulations, you are allowed to reproduce or use those sections of the SPCC plan in the FRP.*

(1) satisfies the appropriate requirements and is equally as stringent;

(2) includes all elements described in the model plan;

(3) is cross-referenced appropriately; and

(4) contains an action plan for use during a discharge.

FRPs must:

•• Be consistent with the National Contingency Plan (NCP) and the Area Contingency Plans.

> *The NCP, also called the National Oil and Hazardous Substances Pollution Contingency Plan, is the federal plan for responding to both oil spills and hazardous substance releases. See **http://www.epa.gov/ oilspill/ncp** for more information.*

•• Identify a qualified individual having full authority to implement removal actions, and require immediate communication between that person and the appropriate federal authorities and responders.

• Identify and ensure availability of resources to remove, to the maximum extent practicable, a worst-case discharge.

•• Describe training, testing, unannounced drills, and response actions of persons at the facility.

•• Be updated periodically.

•• Be submitted for approval with each significant change.

Deadlines for Preparing and Submitting FRPs[7]

The time that you have to prepare and submit a FRP will vary depending on several factors, including the following:

- *Notification from EPA Regional Administrator:* If EPA notifies your facility that you are required to submit a facility response plan, you must prepare and submit a plan within **six (6) months**.

- *Newly Constructed Facilities*: If your facility is newly constructed, you are required to submit the FRP **prior to the start of operations**. After sixty (60) days, you must make adjustments to the FRP to reflect changes that occur during the startup phase and resubmit the FRP.

- *Planned Facility Changes*: If your facility undergoes a planned change in design, construction, operation, or maintenance that places it in the designation of a **substantial harm facility**, then you must submit a FRP **prior to the start of operations** of the portion of your facility undergoing the changes.

- *Unplanned Facility Changes*: If your facility falls under the substantial harm facility designation because of an unplanned event or change in characteristics, then you must submit your FRP **within six (6) months of the unplanned event**.

Response Plan Maintenance

Under 40 CFR 112.20(g), facilities must periodically review their response plans to ensure consistency with the National Oil and Hazardous Substances Pollution Contingency Plan (NCP) and Area Contingency Plans (ACPs). Consequently, a facility that is required to prepare a response plan must review relevant portions of the NCP and the applicable ACPs annually and update its FRP as appropriate. You must submit revised portions of your response plan within 60 days of each facility change, that may materially affect (1) the response to a worst case discharge or (2) the implementation of the response plan.

> *Area Contingency Plans (ACPs)* include detailed information about resources (e.g., equipment and trained response personnel) available from the government agencies in the area. They also describe the roles and responsibilities of each responding agency during a spill incident. You can order copies of ACPs from the National Technical Information Service (NTIS) by calling 1-800-553-6847. To obtain the NTIS ordering number for your area's ACP, first call the RCRA/UST, Superfund and EPCRA Hotline at 1-800-424-9346 or 703-412-9810.

---

[7] The initial statutory deadline for "substantial harm facilities" **either** to submit FRPS **or** to stop handling, storing or transporting oil was February 18, 1993. EPA's regulatory deadline for "substantial harm facilities" and "significant and substantial harm facilities" to submit FRPs or stop handling, storing or transporting oil was August 30, 1994, the effective date of the FRP rule.

Recordkeeping

FRP Requirements Not Applicable: If you determine that the response planning requirements at 40 CFR 112.20 do not apply to your facility, then you must certify and maintain a record of this determination using 40 CFR 112, Appendix C, Attachment C-II.

FRP Requirements Applicable: If your facility is subject to the response planning requirements at 40 CFR 112.20, you are required to maintain the response plan at the facility. You are also required to maintain updates to the plan to reflect material changes to the facility and to log activities such as discharge prevention meetings, response training drills, and exercises. You must keep the records of these activities for a period of five years.

Training and Response Drills

All facilities (i.e., "substantial harm" and "significant and substantial harm" facilities) subject to facility response planning requirements must address training and response drills. Oil spill response training is an important element in EPA's oil spill prevention and preparedness efforts. Because operator error is often the cause of an oil spill, training and briefings are critical for prevention of a spill as well as response to a spill. Training encourages up-to-date planning for the control of, and response to, an oil spill and also helps to sharpen operating and response skills, introduces the latest ideas and techniques, and promotes interaction with the emergency response organization and familiarity with the facility's SPCC and FRP plans.

Under 40 CFR 112.20(h)(8), FRPs must include:

- Information about self-inspection drills, exercises, and response training, including descriptions and logs of training and drill or exercise program; and

- Documentation of tank inspections, equipment inspections, response training meetings, response training sessions, and drills and exercises.

Consequently, FRPs may be revised based on evaluations of the drills and exercises.

Facility Response Training Programs.
Under 40 CFR 112.21 of the Oil Pollution Prevention regulation, facilities are required to develop and implement facility response training programs. It is recommended that the training program be based on the **USCG Training Elements for Oil Spill Response**, as applicable to facility operations. An alternative program can also be acceptable if approved by the Regional Administrator.

> *The PREP guidelines (USCG-X0191) and the Training Reference for Oil Spill Response (USCG-X0188) are available by mail or fax:*
> *TASC Department Warehouse*
> *3341Q 75th Avenue*
> *Landover, MD 20785*
> *FAX: (301) 386-5395*
> *When requesting copies, please indicate the document name and publication number.*

Facility Response Drills/Exercises. Under 40 CFR 112.21, your facility is also required to develop and implement a program of response drills and exercises, including evaluation procedures to test the effectiveness of your response plan. A program that follows the National Preparedness for Response Exercise Program (PREP) will meet EPA's exercise requirements. An alternative program can also be acceptable if approved by the EPA RA.

4.6.4 Oil Spill Notification and Response

Notification - The "One" Immediate Phone Call to the NRC

NATIONAL RESPONSE CENTER

1-800-424-8802

In the Washington, D.C. area:

703-412-9810

**For more information on the NRC, access
http://www.epa.gov/oilspill/NRC**

- • Immediately notify the National Response Center (NRC) of discharges/ releases of oils and hazardous substances by calling the NRC number.

- • If notifying the NRC is not practicable, then immediately notify the pre-designated On-Scene Coordinator of EPA or the USCG. (This means that you must know who your designated On-Scene Coordinator is before the release or discharge occurs.)

- • As required by the relevant Area Contingency Plan, report spills to the state, the tribal government, the territory or commonwealth where the spill occurred.

When an oil spill enters into or threatens any navigable water in the United States, coordinated teams of local, state, and national personnel are called upon to help contain the spill, clean it up, and assure that damage to human health and the environment is minimized. EPA has established requirements for reporting spills into navigable waters or adjoining shorelines. Specifically, facilities are required to report discharges of oil in quantities that may be harmful to public health or welfare or the environment.

EPA has determined that discharges of oil in quantities that may be harmful include those that:

- •• Violate applicable water quality standards;

- • Cause a film or "sheen" upon or discoloration of the surface of the water or adjoining shorelines.

- Cause a sludge or emulsion to be deposited beneath the surface of the water or upon adjoining shorelines.

Also see related notification and reporting requirements for the discharge/release/spill of hazardous substances under EPCRA and CERCLA summarized in this guide in Section 7.3 *Emergency Release Notification.* For a quick, multimedia reference guide to the notification requirements under CWA, OPA, RCRA, EPCRA, CERCLA, CAA and OSHA, see Section 9.3 *Notification and Response Requirements*

Any person in charge of a vessel or onshore or offshore facility should notify **the National Response Center (NRC) at 1-800-424-8802 or 703-412-9810 (Washington, D.C. area)** as soon as he/she had knowledge of a discharge from a vessel or facility. Spills or releases of

> *You should also be aware of state, tribal, and local requirements for spill reporting. For example, there may be a requirement to report all spills meeting certain quantity thresholds even if the spill does not leave the contained area.*

oil which reach navigable waters or adjoining shoreline (including storm drains) or land areas which may threaten waterways must always be reported to the NRC.

Reporting to the National Response Center

When you contact the National Response Center, the staff person will ask you for the following information:

- • Your name, location, organization, and telephone number.
- • Name and address of the party responsible for the incident.
- • Date and time of the incident.
- • Location of the incident.
- • Source and cause of the release or spill.
- • Types of material(s) released or spilled.
- • Quantity of materials released or spilled.
- • Danger or threat posed by the release or spill.
- • Number and types of injuries.
- • Weather conditions at the incident location.
- • Any other information that may help emergency personnel respond to the incident.

The NRC records and maintains all spill reports in a computer database called the Emergency Response Notification System (ERNS), which is available to the public. The NRC relays the spill information to the EPA and U.S. Coast Guard (USCG), depending on the location of the incident.

Specifically, the NRC notifies representatives of EPA or the USCG, known as On-Scene Coordinators (OSCs) are notified. The OSC is the federal official charged with directing a spill response through the Unified Command/Integrated Command System adopted by EPA and USCG. This intergovernmental coordinating system encourages, wherever possible, shared decision making by the federal lead response agency (EPA or USCG), the state(s) and the party responsible for the discharge/release.

Spill Response

The first and most immediate response is that of your facility. For this reason, the quantity, operation, and location of your facility's response equipment and supplies are all critical to effective oil recovery.

•• **SPCC/FRP Regulated Facilities (or Substantial Harm Facilities):** Within the SPCC-regulated community, facilities that may cause substantial harm to the environment or exclusive economic zone based on the quantity and location of their oil storage are required to prepare FRPs to ensure that these facilities have the capability to response to worst case scenario discharges. FRPs greatly assist the facility and response agencies to expedite and coordinate cleanup efforts.

• **Other SPCC Regulated Facilities:** It is recommended that all other facilities in the SPCC-regulated community be prepared to respond to a spill by identifying control and response measures in their SPCC Plans. Every facility should have appropriate spill response equipment available and easily accessible. Absorbent pads and booms, disposal containers or bags, shovels, an emergency response guidebook, and a fire extinguisher are essential components of a spill kit. Portable pumps may also be a good investment. It is also recommended that facilities coordinate with local responders, other nearby facilities, and contractors before a spill occurs so that response is accomplished most efficiently. Facility personnel, including seasonal employees, must be educated and trained in spill response, notification, and oil recovery. By being prepared to respond, the impact of a discharge on human health or the environment may be minimized and cleanup costs and fines resulting from improper notification or response reduced.

• **First Response:** In the event of an oil spill, the facility response plan is immediately activated. Depending on the nature of the spill, local, area, regional, or national plans may also be activated. The OSC will activate these plans if the facility is not equipped or capable of handling the response.

• **On-Scene Coordinators:** The designated OSC from EPA or USCG is responsible for determining how to respond to the spill, i.e., determining the resources, both personnel and equipment needed. The OSC does this based on his/her assessment of several factors, including the following: the magnitude and complexity of the spill; the availability of appropriate response equipment and trained personnel; and the ability of the responsible party, or local and/or state responders to respond to the spill.

   Although the OSC is responsible for coordinating federal efforts with local, state and regional response efforts, in practice the role of the OSC varies. Depending on the OSC's assessment, he/she may do the following: direct the response; direct the response in cooperation with other parties; oversee that the responses is conducted by other parties; provide limited or periodic oversight; or determine that a federal response is not needed.

For example, small spills may be cleaned up by the facility (or responsible party) or by local response agencies, while larger spills may require regional response efforts. In either cases, the OSC is required to oversee and monitor the spill response to make sure that all appropriate actions to prevent threats to human health or the environmental are taken. If, however, a facility is handling a smaller spill adequately, the OSC may not go to the site.

- *Oil Recovery:* The OSC, response teams, and a network of experienced agencies will decide on the most effective method of cleanup. These agencies must coordinate cleanup efforts carefully and efficiently to protect response personnel, recreational areas, drinking water reservoirs, and wildlife from the potentially catastrophic effects of an oil spill.

  Selecting the best method, or combination of methods, for recovering oil after a spill is based on several factors. The type and amount of oil spilled and the water body are the most important considerations. The mechanisms most frequently employed to control oil spills and minimize their impact on human health and the environment fall into four broad categories: (1) mechanical containment or recovery includes booms, barriers, skimmers, and sorbent materials; (2) chemical and biological methods include dispersants, gelling agents, and biological agents; (3) physical methods include wiping, pressure washing, raking, and bulldozing, also scare tactics, such as floating dummies, to keep birds away from a spill area; and (4) natural processes, which include evaporation, oxidation, and biodegradation.

For more information, visit EPA's Oil Program at **http://www.epa.gov/oilspill/**.

4.7 Compliance Issues For Selected Activities

4.7.1 Land Application of Wastewater

Land application. Land application is the process of discharging wastewater from an industrial facility, such as a food processing facility, to land or agricultural crops. This process can be beneficial to the crops which utilize the water and the carbohydrates and nutrients in the wastewater. Land application generally is regulated by the state and may require a permit. The permit is designed to regulate contaminants in the wastewater, and ensure that the wastewater does not run off into nearby waterways.

Some typical requirements that may be included in a permit for the land application of wastewater include:

- A "no discharge" requirement prohibiting runoff to waterways

- Prohibitions of land application (including spraying)during wet weather or when the ground is frozen

- Monitoring of pollutant levels in the wastewater or sludge

- Limits on the amount of pollutants and the amount of wastewater applied.

- Installation of monitoring wells and monitoring of groundwater

- Installation of a pretreatment system to pretreat the wastewater before land application.

Check with your state regulatory agency for more information on the requirements for land application of industrial wastewater.

Overland flow treatment system. Some facilities may use an overland flow treatment system for treating their wastewater. This type of system, which results in discharges to a receiving water, requires a NPDES permit. Check with EPA or your state regulatory agency for more information.

4.7.2 Construction or Plant Modification Activities

There are other federal regulations that potentially apply to construction or plant modification activities at your facility, including regulations addressing wetlands and endangered species. These are discussed below.

Wetlands. Activities, such as construction or plant modification, at your facility that impact wetlands may require a permit. Wetlands, which commonly are called swamps, marshes, fens, bogs, vernal pools, playas, and prairie potholes, are a subset of "waters of the United States" as defined in Section 404 of the CWA. The placement of dredge and fill material into wetlands and other water bodies (i.e., waters of the United States) is regulated by the U.S. Army Corps of Engineers (Corps) under 33 CFR 328. The Corps regulates wetlands by administering the CWA Section 404 permit program for activities that impact wetlands. EPA's authority under Section 404 includes veto power of Corps permits, authority to interpret statutory exemptions and jurisdiction, enforcement actions, and delegating the Section 404 program to the states.

Because wetlands and the regulations protecting them are dynamic, it is important to check with the Corps district office even if you think a permit may not be required for your activity. If your project area includes wetlands, the Corps district office also may suggest that your facility retain a consultant to delineate wetland boundaries. In addition to conducting the wetland delineation, some wetland consultants also can help with the permit application process.

In addition to federal regulations, some state and local governments have laws protecting wetlands. Laws may include those that require permits for construction in wetlands. To find out if your proposed activities require a state permit, contact the appropriate department (e.g., state department of water resources, natural resources, or the environment) in the state where the activities will take place.

> *Some states or local governments may have stricter wetland regulations than CWA Section 404, so if your activity does not require a Section 404 permit (and involves a wetland) you still should consult with the appropriate state agency.*

Endangered or Threatened Species. The federal Endangered Species Act (ESA), administered by the U.S. Department of Interior's (DOI) Fish and Wildlife Service (USFWS) and the Department of Commerce's National Marine Fisheries Service (NMFS), establishes a program for the conservation of endangered and threatened species and the habitats in which they are found. State laws or regulations may be more, but not less restictive, than the federal ESA or its regulations.

If you are engaged in, or planning to engage in, activities such as construction or plant modification, you must be aware if any endangered or threatened species exist on the property involved, or if the property is considered part of a listed species' critical habitat. If neither is the case, the ESA does not apply. However, if the action will "take" a species or degrade critical habitat, some form of mitigating action must be taken to prevent harming the species. Contact your local USFWS endangered species coordinator or talk with a qualified consultant to clear up any specific questions relating to your facility's activities.

> *The term "take" includes harassing, harming, hunting, killing, capturing, and collecting.*

SECTION 5 CONTENTS

5. HOW DO I COMPLY WITH SAFE DRINKING WATER REGULATIONS?

As a food processing facility, you are responsible for the drinking water supplied to your employees and the water used in food processing operations. The Safe Drinking Water Act (SDWA) protects the water supply through water quality regulations and source protection, such as underground injection control (UIC) regulations. SDWA requirements apply to all public water systems (PWSs). Therefore, it is important that you determine whether you are a PWS, and if so, which SDWA requirements apply to your water system. This section explains the SDWA program and how you can comply with the regulations.

5.1 Introduction

The purpose of the SDWA is to protect public health by regulating PWSs and underground injection. EPA is responsible for writing regulations to carry out the provisions of the Act, including drinking water standards, monitoring/reporting, and public notification requirements. The entities that supply public water are responsible for making sure that the water meets EPA's standards.

SDWA 1996 Amendments. Since being passed by Congress in 1974, the SDWA has been amended twice, most recently in 1996. The 1996 SDWA Amendments provide:

(1) **New and stronger approaches to prevent contamination of drinking water.** The Amendments established a strong new emphasis on preventing contamination problems through source water protection, capacity development, and operator certification programs.

(2) **Better information for consumers, including "right to know."** The Amendments specify that the public be provided with or given access to data collected, analyses done or implementation strategies developed under new SDWA programs through consumer confidence reports and other provisions for improved consumer information.

(3) **Regulatory improvements.** Regulatory improvements by EPA include: (a) new risk-based contaminant selection; (b) cost-benefit analysis and research for new standards; (c) small system technologies, variances, and exemptions; (d) extension of compliance time frames; (e) monitoring reforms; and (f) streamlining of enforcement processes.

(4) **New funding for states and communities through a Drinking Water State Revolving Fund.** One of the most notable features of the new law is the authorization for states to use State Revolving Funds (SRFs) for new prevention programs, such as source water protection, capacity development, and operator certification programs, as well as for the State's overall drinking water program.

For more information on the SDWA and its 1996 Amendments, see EPA's Office of Groundwater and Drinking Water website at **http://www.epa.gov/OGWDW/**.

Additional Requirements. In addition to EPA's SDWA requirements, water used in food processing operations must meet the Food and Drug Administration (FDA) and the United States Department of Agriculture (USDA) requirements. The FDA, under its good manufacturing practice regulations, requires that "any water that contacts food or food-contact surfaces shall be safe and of adequate sanitary quality" (Current Good Manufacturing Practice in Manufacturing, Packing, or Holding Human Food, 21 CFR Section 110.37). In addition, the USDA's Food Safety and Inspection Service (FSIS) sets standards for activities associated with the production of meat and poultry products, including standards involving water quality, and water use and reuse.

5.2   How Does the Program Work?

The federal drinking water program may be delegated to states if they meet requirements in the law and the regulations. This is known as primacy. Fifty-four of 56 states and territories have been delegated primacy to run the drinking water program. EPA sets standards and provides guidance, technical assistance, and some financing to these agencies. EPA has primacy in Wyoming, Washington, D.C., and all Tribal Lands, and may also take enforcement action in a primacy state where the state does not take an enforcement action in response to a violation. The primacy agency, whether EPA or a state, is responsible for tracking sample results, conducting detailed inspections called sanitary surveys, and taking enforcement actions such as imposing fines and penalties when necessary.

5.3   How Do I Know If I Am Regulated?

The SDWA regulations apply to you if your facility operates a **public water system (PWS)** or receives water from a PWS and provides treatment to it. Prior to 1996, SDWA defined a PWS as "a system for the provision to the public of **piped water for human consumption** if such system has at least fifteen service connections or regularly serves at least twenty-five individuals." The 1996 Amendments expanded the means of delivering water to include not only pipes, but also **other constructed conveyances**, such as ditches, waterways, flumes, mine drains, and canals. Furthermore, if a **water supplier** provides water to at least twenty-five individuals or fifteen connections at any time on or after August 6, 1998, the supplier is considered a PWS.

Public water systems are divided into three categories:

- ***Community water systems.*** Generally serve the same people year round (e.g., a small town).

- ***Transient non-community systems.*** Serve people only for a portion of the time (e.g., hotels, restaurants, or highway rest stops).

•• ***Non-transient, non-community systems.*** Systems that serve at least 25 people for over six months of the year, but the people generally do not actually live at the facility (e.g., schools or factories).

To understand whether your facility is a PWS, you should know the following information:

(1) do you or another entity supply the water;
(2) the number of people served by your system; and
(3) their pattern of water use.

Your facility is most likely operating a non-transient, non-community water system if your sole source of drinking water is **not** from a municipal or district water supplier.

All PWSs must meet the national primary drinking water regulations (see below). In addition, some states regulate **smaller** facilities than EPA. Contact your state/territory/Tribal government to determine if any additional requirements that apply to your PWS.

> *Note: Even if a state/territory/Tribal government does not have primacy for the SDWA program, they may have additional requirements and should be contacted.*

5.4 What Are The National Drinking Water Regulations?

Under the SDWA, EPA establishes national primary and secondary drinking water regulations designed to protect the public health and the aesthetic quality of the water.

•• ***Primary drinking water regulations*** are health-based and enforceable.

• ***Secondary drinking water regulations*** are based on the aesthetic quality of the water and are non-enforceable guidelines.

Remember that states have the option to set drinking water standards that are more stringent than those set by EPA. Contact your state regulatory agency to determine if any additional state requirements apply to your water system.

5.4.1 National Primary Drinking Water Regulations

Maximum Contaminant Levels and Treatment Techniques

EPA has established national primary drinking water regulations (NPDWRs). As part of the NPDWRs, EPA has developed **maximum contaminant levels (MCLs)** and/or **treatment**

techniques (TTs) for **more than 80** contaminants, including organics, inorganics, radionuclides, and microbiologicals.

MCLS are drinking water standards that are based on maximum contaminant level goals (MCLGs) and other factors. MCLs are in effect for 72 contaminants. An MCL defines the amount of that contaminant allowed to be present in the drinking water. MCLs are set based on known or anticipated human health effects, and the ability of various technologies to remove the contaminants, their effectiveness, and cost of treatment. MCLs are **enforceable** standards, and therefore, are the levels against which the water samples from regulated systems are judged for compliance with the regulations. To comply with MCLs, your facility may use any state-approved treatment.

When it is not economically or technologically feasible to set an MCL for a contaminant (e.g., when the contaminant cannot be measured easily), EPA may require use of a particular **treatment technique (TT)** instead. A TT is an **enforceable** procedure or level of technological performance that PWSs must follow to ensure control of a contaminant. TTs are set for nine contaminants. The technique specifies the design for part of the drinking water treatment process (such as filtration or corrosion control) to remove these contaminants and prevent health problems. Examples of two TT rules include the following:

- **Lead and Copper Rule.** The Lead and Copper Rule (40 CFR 141, Subpart I) is a set of treatment technique requirements. If you operate a community system or non-transient, non-community water system, you must comply with these requirements. The rule requires all systems which do not meet the specified lead and copper action levels at the tap to optimize corrosion control treatment in an effort to minimize the levels of these contaminants. The rule has five major components: (1) monitoring, (2) distribution system corrosion control, (3) source water treatment, (4) public education, and (5) lead service line replacement. Each of these components can be considerably complex and you should work closely with your primacy state to determine the exact requirements that apply to your system.

- **Surface Water Treatment Rule (SWTR).** The SWTR (40 CFR 141, Subpart H), promulgated in 1989, applies to all PWSs using surfacewater sources or ground water sources under the direct influence of surface water. It includes treatment technique requirements for filtered and unfiltered systems that are intended to protect against the adverse health effects of exposure to *Giardia lamblia, Legionella*, as well as many other pathogenic organisms. To comply with the monitoring provisions of SWTR, PSWs must conduct analyses of total coliforms, fecal coliforms, heterotrophic bacteria, turbidity, and temperature, as well as measure residual disinfectant concentrations.

Contact EPA's Safe Drinking Water Hotline at 1-800-426-4791 or see EPA's website at **http://www.epa.gov/ogwd000/methods/swtr_tbl.html** for more information.

Monitoring and Reporting

As a supplier of water, you must collect samples from your water system, submit them to an EPA or state-approved laboratory (sometimes known as a

Some states perform the sampling for the regulated systems in their state. You must contact your state (or other primacy agency) to find out if this applies to your system.

certified laboratory) for analysis, and send the analytical results to the regulatory agency (usually state or county health department). New PWSs may have to perform initial monitoring more frequently. The type of analysis performed, the sampling frequency, and the location of the sampling point vary from system to system and chemical to chemical. Sampling requirements for all systems can be found in 40 CFR 141.21, 141.22-24, 141.26, 141.35-42, 141.60, 141.74, 141.80, 141.83, and 141.86-88.

Monitoring reports are required to be sent to the regulatory agency (usually state or county health department). These reports must include:

> *Note: Waivers may be available from the state to reduce monitoring requirements for some contaminants.*

- •• Date, place, and time of sampling and name of the person who collected the sample(s)

- •• Identification of the sample (e.g., routine or check sample, raw or treated water)

- • Date of analysis, laboratory conducting analysis, name of person responsible for analysis, and analytical method used

- •• Analytical results.

Reporting requirements are found in 40 CFR 141.31-33. 141.35, 141.75, 141.90 and 143.5. If a problem is detected through sampling, there are immediate retesting requirements that go into effect and strict instructions for reporting about the problem.

Public Notification and Violations

In addition to reporting to the primacy agency, you must notify the public if there is a violation. The timeframes and methods of public notification differ depending on the kind of violation(s) (e.g., those with acute health risk, non-acute health risk, or other kinds) (40 CFR 141.32). Methods of public notification could include one or more of the following:

> *EPA is revising public notification regulations and is scheduled to propose them in the Fall 1998.*

direct mail; local newspaper; local radio and/or television; hand delivery; or continuous posting in conspicuous places. **Note: In the event of a violation, you must keep monitoring as required by the rules.**

Violations are divided into two categories: Tier 1 and Tier 2, depending on the seriousness of the violation.

> **Tier 1 violations.** Violations of a drinking water standard require prompt notification. (Times vary from "as soon as possible" for acute health hazards to within 45 days for chronic hazards, and are also based on communication mechanisms available to the supplier of drinking water.)

> **Tier 2 violations.** Violations related to monitoring, reporting, or recordkeeping must be reported within three months. In addition to notification when there is a violation, **a special one-time notification is required concerning the contaminant lead**. The lead

notification should have been made by June 19, 1988. If you have not made the notification yet, contact your primacy agency for assistance.

Recordkeeping

If regulated, your food processing facility must maintain certain records for required periods of time. These records and time periods are specified in 40 CFR 141.33, 141.75, 141.80 and 141.91. Depending on the types of records, required time periods range from three (3) to ten (10) years.

5.4.2   National Secondary Drinking Water Regulations

National secondary drinking water regulations (NSDWRs) are federal guidelines regarding taste, odor, color, and certain other non-aesthetic effects of drinking water. As part of the NSDWRs, EPA developed secondary MCLs for 15 contaminants. Additional guidelines under the NSDWRs include those for monitoring, analytical methods, and public notification.

These regulations are **not** federally enforceable. EPA recommends them to states as reasonable goals, but federal law does not require public water systems to comply with them. States may however, adopt their own **enforceable** regulations governing these concerns. To be safe, check your state's drinking water regulations and contact your primacy agency.

5.5   Underground Injection Control (UIC) Requirements

The SDWA UIC program (40 CFR 144-48) is a permit program designed to protect underground sources of drinking water by regulating the injection of **liquid waste** into five classes (I through V) of injection wells. EPA may delegate enforcement of UIC requirements to primacy states. However, EPA maintains primacy enforcement authority for all wells in 13 states and territories, all Tribal Lands, and for some classes of wells in 7 states.

> *Note: Even if a state/territory/Tribal government does not have primacy for the UIC program, they may have additional requirements and should be contacted. In addition, there also may also be local requirements (e.g., county health department, building code requirements, etc.).*

Your facility is subject to these regulations only if it injects liquid waste underground, and only under the following conditions:

- **Either** your facility maintains a well (or hole) that is deeper than its largest surface dimension, where the principal function of the hole is emplacement of fluids;

•• **Or** your facility disposes of non-domestic waste/wastewater (such as laboratory waste, industrial waste, storm water) into a subsurface disposal system such as a septic system, drainage well, drywell, or cesspool;

• **Or** your septic system or cesspool is used solely for the disposal of sanitary waste and has the capacity to serve more than 20 persons per day.

If your facility meets any one of these criteria, you are required to obtain UIC authorization by permit or by rule from your primacy agency to inject liquid waste. UIC permits include design, operating, inspection, closure, and monitoring requirements. Furthermore, wells used to inject hazardous wastes also must comply with RCRA corrective action standards in order to be granted a RCRA permit, and must meet applicable RCRA land disposal restriction (LDR) standards. See Section 8.0 *How Do I Comply With the Hazardous Waste Regulations?* for more information on LDR standards.

If your facility disposes of liquid waste by injection, it is most likely to a Class V well. These wells are currently authorized by rule, which means they do **not** require a permit if they do not endanger underground sources of drinking water (USDWs) and meet certain minimum requirements. Under the conditions of the UIC regulation, you are required to submit to the primacy agency basic inventory information about the Class V injection well, and ensure that the well is constructed, operated, and closed in a manner which protects USDWs. The primacy agency may request additional information or require a permit in order to ensure groundwater quality is adequately protected. Furthermore, many primacy state programs have additional prohibitions or permitting requirements for certain types of Class V injection wells.

> *Class V wells include shallow nonhazardous industrial waste injection wells, septic systems, and storm water drainage wells.*

> *If any amount of hazardous waste is discharged to a Class V well, you must immediately notify the your primacy agency.*

On July 29, 1998, EPA **proposed a rule**, 40 CFR Part 144, Subpart G - *Requirements for Owners and Operators of Class V Injection Wells* (63 FR 40586), which focuses on high-risk Class V injection wells in source water protection areas (SWPAs) that are known to pose the greatest threat to USDWs. These high-risk wells include motor vehicle waste disposal wells, industrial waste disposal wells, and large-capacity cesspools. The proposed regulation would affect the owners and operators of these wells in SWPAs delineated for community water systems and non-transient, non-community water systems that rely on at least one groundwater source. For more information, contact EPA's Safe Drinking Water Hotline at 1-800-426-4791.

SECTION 6 CONTENTS

6. HOW DO I COMPLY WITH AIR REGULATIONS?

6.1 Introduction

This section presents an overview of the Clean Air Act (CAA) and a discussion of the common air emissions from food processing facilities. Although total air emissions by the food processing industry typically are less than other manufacturing industries, some sources may emit sufficient air pollution to be regulated under CAA. This section identifies some common types of air emissions produced by food processing facilities; the federal standards that apply to those emissions; how to calculate your facility's total emissions; how to determine whether your facility meets federal thresholds for regulations; and discusses when you need an air permit.

6.2 What is the Clean Air Act?

The federal CAA and the Clean Air Act Amendments (CAAA) of 1990 regulate air pollution in the United States. The CAA authorizes EPA to codify rules and regulations that will ensure that the public and the environment will be protected. Although the CAA is a federal law, state and local air pollution control agencies do much of the work in carrying out the act. It is important for you to know all of the applicable federal, state, and local regulations, because in many instances, state and local regulations may be more stringent than the federal regulations and/or include additional requirements.

The CAA and the CAAA of 1990 can be characterized in terms of three programs: (1) air quality regulations; (2) new source performance standards; and (3) specific pollution problems (e.g., hazardous air pollutant emissions).

Air Quality Regulations

Pursuant to Title I of the Clean Air Act, EPA has established National Ambient Air Quality Standards (NAAQS) for six criteria air pollutants: ozone, sulfur oxides (SOx), carbon monoxide (CO), nitrogen dioxide (NO_2), particulate matter (PM), and lead (Pb) (see 40 CFR 50).

> **How NAAQS for Criteria Pollutants May Affect Food Processing Industries**
> SO_2, NO_x, and **PM** result from the combustion of fossil fuels (e.g., industrial boilers, see Section 6.5.1 Boilers or Steam Generating Units). Some significant sources of **particulate emissions** result mainly from solids handling, solid size reduction, cleaning, roasting, drying, and cooking (e.g., PM_{10} results from flour, sugar, and other dry ingredients). Some of the particulates are dusts, but others are produced by the condensation of vapors ranging in the low-micrometer or submicrometer particle-size. **VOC** emissions may result from fryers (e.g., doughnuts, french fries), direct use of ethanol, by-products of yeast fermentation (ethanol), and from lubricating oils for machinery.

Because EPA cannot directly regulate ozone, volatile organic compounds (VOCs) and nitrogen oxides (NOx), which significantly contribute to the formation of ground level ozone, are regulated by EPA. Sulfur oxides are measured in the ambient air as sulfur dioxide (SO_2).

EPA has developed two types of standards for these criteria pollutants: the **primary standard** protects health, whereas the **secondary standard** is intended to protect environmental and property damage. A geographic area that **meets or does better** than the primary standard is called an **attainment area**; areas that **do not meet** the primary standard are called **nonattainment areas**. Many urban areas are classified as nonattainment for at least one criteria air pollutant. Nonattainment areas are designated further as marginal, moderate, serious, severe, or extreme, depending on the amount of effort needed in the nonattainment area to achieve NAAQS. Failure of a state to meet attainment deadlines results in reclassification of a nonattainment area to the next higher classification with more stringent control requirements.

New Air Quality Standards for Ozone and Particulates. In July 1997, EPA promulgated new standards (i.e., NAAQSs) for ozone and particulate matter. As a result of the new standards, additional areas of the country may be designated as nonattainment for ozone and particulate matter. Check with your state and local air pollution control authorities to find out if these new standards affect your facility. For a list of sources and maps of nonattainment areas, refer to the EPA's AIRSWeb site at **http://www.epa.gov/airprogm/airs/web/maps.htm/**.

State Implementation Plans. Under Title I of the CAA, all states must prepare a **state implementation plan** (SIP) for achieving attainment by a specified date (see CAA Section 110, 42 U.S.C. 7410). While EPA promulgates rules and regulations that limit emissions from specific types of facilities and for specific air pollutants, each state must promulgate appropriate rules and regulations through the SIP process, depending on the state's attainment status. EPA must approve each SIP, and if the SIP is not acceptable, EPA can take over enforcing the CAA (e.g., permitting authority) in that state (see CAA Section 110(a)(2), 42 U.S.C. 7410).

> *Under Section 110 of the CAA, states are required to implement **new source review (NSR)** provisions for nonattainment areas in their SIPs. Your facility is subject to new source review requirements if you are a new major source or an existing major source with significant modifications to equipment at your facility.*
>
> *A parallel program that applies to attainment areas is the **Prevention of Significant Deterioration (PSD)** which pertains to certain types of stationary sources that have the potential to emit more than 100 tons per year of any regulated pollutant or any source that emits more than 250 tons per year of any one pollutant. For more information on these two programs, see Section 6.3.3 Air Pollution Permits.*

Existing sources located in nonattainment areas may be required to install **reasonably available control technology (RACT)**. RACT is defined as "devices, systems, process modifications, or other apparatus or techniques that are reasonably available" in order to obtain attainment status (CAA Section 172). EPA has established RACT guidelines for over 30 major source categories of nonattainment pollutants and has guidelines under development for additional categories. These guidelines are implemented by the states through their SIPs. Although EPA has not prepared RACT guidelines for food processes, such as vinegar generation, a state may develop RACT guidelines or require major sources to prepare their own RACT requirements. Therefore, RACT guidelines may vary greatly from state to state: contact your state permitting authority to find out what RACT guidelines apply to you.

Under the CAAA, existing sources must install RACT for VOCs and NOx. In addition, states with ozone nonattainment areas must revise their SIPs to address various new requirements, including incremental reductions of VOCs.

New Source Performance Standards (NSPS)

Section 111 of the CAA required EPA to identify categories of new and modified sources that contribute significantly to air pollution and endanger public health or welfare. After identifying approximately 60 source categories (e.g., grain elevators, fossil fuel-fired generators, steam generating units) that are designated by size as well as type of process, EPA established uniform, national emission standards known as **New Source Performance Standards (NSPS)** in 40 CFR 60. See Section 6.5.1 *Boilers or Steam Generating Units* for more information about requirements for fossil fuel-fired generators and steam generators.

These emission standards for categories of major new, modified, or reconstructed sources are based on the **best available control technology (BACT)**. EPA is required to consider economic, energy, and non-air environmental factors in setting NSPS. Note that the NSPS program sets a minimum level of control for new and modified sources of air pollution. More stringent control may be required under either the **prevention of significant deterioration (PSD)** or the nonattainment pre-construction permitting programs. See Section 6.3.3 *Air Pollution Permits* for more information.

Monitoring, notification, and recordkeeping requirements. Owners and operators of sources subject to NSPS must meet notification and recordkeeping requirements listed at 40 CFR 60.7. They must also meet all monitoring requirements as presented in 40 CFR 60.13, or the applicable subpart.

Specific Pollution Problems

In addition to NAAQS and NSPS, the CAA requires EPA to address specific pollutants, known as hazardous air pollutants (HAPs). HAPs, or air toxics, are chemicals that cause serious health and environmental harm. HAPs are released from stationary sources throughout the country and from motor vehicles.

Under Title III of the CAA, EPA established **National Emission Standards for Hazardous Air Pollutants (NESHAP)**. The list of regulated HAPs can be found in Section 112(b)(1) of the CAA. The CAAA further directed EPA to develop a list of sources that emit any of the HAPS and to develop regulations for these categories of sources. To date, EPA has listed 188 hazardous substances and 174 source categories and has developed a schedule for the establishment of emission standards for these sources. EPA is developing these emission standards for both new and existing sources based on **maximum achievable control technology (MACT)**. MACT is defined as the control technology that achieves the maximum degree of HAP emission reductions, taking cost and other factors into account (see CAA 112(b)).

> *EPA is developing a few **MACT** standards for the food processing industry, such as controls to reduce acetaldehyde, which is produced as a by-product during the fermentation process in the baker's yeast manufacturing industry.*

Monitoring, notification, recordkeeping, and reporting requirements. Notification requirements for NESHAP source categories are listed in 40 CFR 63.9. (Note: EPA is planning to revise the notification requirements.) Monitoring requirements for NESHAP source categories are presented in 40 CFR 63.8 and recordkeeping and reporting requirements are listed in 40 CFR 63.10.

The state in which your food processing facility operates also may regulate HAPs. Check with your state and local air pollution control authorities to find out if additional or more stringent standards for HAPs apply to you.

6.3 What Are My Air Emissions and How Do I Manage Them?

Because your facility emits air pollutants, it is important that you comply with air pollution control requirements and find methods for reducing air emissions from your facility in order to protect yourself, your co-workers, and the quality of air in your community. There are several steps you should follow to responsibly manage air emissions from your food processing facility, including:

- • Identify the products or processes you use that produce air pollutants

- • Calculate all **actual** and **potential** air emissions that your facility emits (This is important for determining whether you are a major or minor source and what federal, state, and local requirements apply to your facility. See Section 6.3.2 *Determining Whether Your Facility Meets Federal Regulations* for more information.)

- • Check with your state and local air pollution control offices and determine which requirements apply to your facility

- • Comply with all applicable regulations, including obtaining the necessary permits. (See Section 6.3.3 *Air Pollution Permits* for more information.)

6.3.1 Identifying and Quantifying Air Emissions

Your food processing facility may be emitting air pollutants (i.e., criteria, HAPs) from both process and ancillary operations, such as refrigeration and steam generation. You should evaluate the processes and ancillary operations at your facility in order to determine the type and amount of pollutants released into the air.

After identifying your facility's air pollutant emissions that are subject to regulation under CAA and state requirements, you are required by the CAA to determine the **actual** amount of air pollutants generated, as well as your facility's **potential to emit** these pollutants. You may need to perform specific calculations to determine your facility's

> *Under Section 112 of the CAA, your facility is required for all regulated pollutants to calculate **actual emissions**, as well as your **potential to emit** these pollutants.*

actual and potential air emissions to determine which threshold for regulation your facility meets. See Section 6.3.2 *Determining Whether Your Facility Meets Federal Regulations* for more information.

You can calculate your facility's actual emissions and potential to emit by one of two ways: (1) pollutant by pollutant; or (2) total of all emissions. Calculating emissions for NOx, SO2, PM generally are done on a pollutant-by-pollutant basis, while total emission calculations for VOCs and HAPs may be done by calculating total pollutant emissions or pollutant-by-pollutant. In order to calculate emissions (actual or potential), you must first determine the following:

- •• The source (e.g., boiler, reactor, etc);
- •• What the source does to cause an emission (e.g., burn fuel, react);
- •• Raw materials or inputs used, and at what rates;
- •• The calculation method (e.g., *AP-42 Compilation of Air Pollutant Emission Factors*, stack test, material balance).

Determining Your Facility's Actual Emissions: Actual emissions can be determined by the following methods: (1) estimates calculated using published emission factors; (2) stack tests; (3) engineering estimates; or (4) material balance methods. In general, facilities may choose which method to use when calculating actual emissions; however, the method chosen is subject to review and approval by the state.

- •• Emission estimates can be calculated using **published emission factors**. Published emission factors are representative values that attempt to relate the quantity of a pollutant released into the atmosphere with the activity associated with the its release. These factors usually are expressed as the weight of pollutant divided by a unit weight, volume, distance, or duration of the activity emitting the pollutant (e.g., pounds of particulate emitted per 1000 gallons of fuel oil burned). Such factors facilitate estimation of emissions from various sources of air pollution. In most cases, these factors simply are averages of all available data of acceptable quality, and generally are assumed to be representative of long-term averages for all facilities in the source category.

  To estimate emissions using published emission factors, use the following general equation:

  $$E = A \times EF \times (1- ER/100)$$

  where E = emissions; A = activity rate (e.g., gallons (in thousands) of fuel oil burned per year) EF = emission factor (e.g., pounds of particulate per 1000 gallons of fuel oil burned); and ER = overal emission reduction efficiency (%).

This general equation for estimating emissions and many emission factors are published in EPA's AP-42 document series entitled, *Compilation of Air Pollutant Emission Factors.* AP-42 emission estimates and factors generally are calculated on a pollutant-by-pollutant basis. The extent of completeness and detail of the emissions information in AP-42 depends upon the availability of published references. Emissions from some processes are better documented than others. AP-42 can be found at EPA's Technology Transfer Network (TTN) website at **http://www.epa.gov/ttn/chief/ap42. html**.

Compilation of Air Pollutant Emission Factors (AP-42)
*Dryers, roasters, ovens, and other equipment used by the food processing industries burn natural gas. Emissions from the combustion of natural gas include nitrogen oxides (NOx), carbon monoxide (CO), carbon dioxide (CO_2), methane (CH_4), nitrous oxide (N_2O), and VOCs, as well as some sulfur dioxide (SO_2) and particulate matter. **AP42 Natural Gas Combustion** is a good reference for calculating these emissions.*

*Fryers (doughnuts, french fries, corn chips) emit particulate matter and small amounts of VOC from the deep fat frying process. See **AP42 Snack Chip Deep Fat Frying** for calculating these emissions.*

•• **Stack tests** can be done to measure short term (e.g., hourly) actual emissions at a maximum production rate. EPA prescribes test methods to measure pollutant emissions, and these are listed in 40 CFR 60, Appendix A. It is likely that your facility is required to do stack testing in order to show compliance with NSPS and NESHAP standards as discussed earlier in this section. Many facilities voluntarily do stack testing if an emission estimate (see below) is not available, or if it is believed that an emission estimate is overestimating your emissions. For example, if AP-42 determines that you are a **major source**, you may want to use stack tests in order to show that your emissions are actually lower than the major source category. By demonstrating that your emissions are too low to be declared a major source, your facility may save time and money spent on permitting fees, pollution control equipment, and other regulatory requirements. See Section 6.3.3 *Air Pollution Permits* for more information.

•• **Engineering estimates** use equipment-specific calculations to determine actual emissions, such as mass transfer calculations, heat transfer calculations, and distillation calculations, among others. This type of emission estimation procedure requires an intimate knowledge of the specific process that generates the emission; the thermodynamic, physical, and chemical properties of the materials involved; and experience in the application of the appropriate calculation equations. The desired results from engineering estimates normally are air pollutant rates per unit time (e.g., lb/hr) emitted from the process or piece of equipment. When using engineering estimates, you can calculate total pollutant emissions or pollutant-by-pollutant.

•• When emission factors are not available and engineering estimates are not practical, the **material balance method** may be used for

---

Ethanol is often used to shine jellybeans. To use **material balance method**, subtract the amount of ethanol emitted as VOCs from the amount of ethanol applied initially.

---

determining your emission rates. The basic concept underlying the material balance method is that the amount of material entering a process (like cooling or preserving) is equal to the amount exiting the process. Therefore, what you purchase as raw material must become part of the finished product, be emitted to the air, released into water, be disposed of as waste, or be accumulated in the inventory. When using the material balance method, you can calculate pollutants by either calculating total emissions, or by calculating pollutant-by-pollutant. This method may be preferable for some businesses that find the options discussed above to be too costly or otherwise impractical.

Determining Your Facility's Potential to Emit. A facility's **potential to emit** is defined as the maximum capacity of a stationary source to emit a pollutant under its physical and operational design. Any physical or operational limitation on the capacity of the source to emit a pollutant, including air pollution control equipment and restrictions on hours of operation or the type or amount of material combusted, stored, or processed, shall be treated as part of its design if the limitation or the effect it would have on emissions is federally enforceable (40 CFR 52.21).

To determine your facility's potential emissions, you can use the following calculation:

Potential emissions (lbs/year) = Lesser value of uncontrolled or allowable emission rate at maximum capacity x 8760 hours

Please keep in mind that your **potential to emit** must account for emissions that **could have** come from any unused equipment, even if these emissions were not included when determining the facility's actual emissions. For example, if your facility has a boiler that operates 24 hours a day for only 300 days a year, you still must calculate your **potential to emit** on the assumption that your boiler operates all 365 days a year.

6.3.2   Determining Whether Your Facility Meets Federal Regulations

A facility's potential to emit pollutants is important in determining how your facility is regulated under CAA, and whether you must obtain a CAA Title V operating permit. Whether and how you are regulated depends on several factors including whether your facility is located in a non-attainment area for a particular criteria pollutant and whether your facility's potential to emit meets the threshold for regulation as a major source. Together these will determine whether you must obtain a Title V permit.

Under the CAA, facilities are classified as major or minor sources based on potential to emit. Generally, a facility is considered a **major source** if its potential to emit is 100 tons per year (tpy) of any criteria pollutant. For facilities in nonattainment areas, the emission rate threshold for major sources varies by pollutant and area classification (e.g., moderate, serious, severe). The following table summarizes these thresholds. Note: The threshold values decrease as the degree of non-attainment increases from marginal or moderate to serious to severe, etc.

> **Note**: a facility can be a major source for more than one pollutant.

A facility is also a **major source** if it has the potential to emit 10 or more tpy of any single HAP, or 25 tpy or more of any combination of HAP emissions.

Based on the above discussion, if your facility meets any of the thresholds shown in Table 6-1, your facility is classified as a **major source**, and you must follow the requirements listed below:

- • Obtain a Title V operating permit (see Section 6.3.3 *Air Pollution Permits*).

- If you are a major source in a nonattainment area, you must reduce your emissions through the use of Reasonably Available Control Technology (RACT) (see Section 6.2 *What is the Clean Air Act?*).

- If you emit HAPs, such as VOCs, check with your state environmental agency because it may regulate other pollutants in addition to those on the federal HAPs list.

Table 6-1. Major Source Emission Rate Thresholds in Nonattainment Areas

Pollutant	Area Classification[1]	Threshold[2]
Ozone	Marginal or Moderate	100 tpy VOCs or NOx
	Serious	50 tpy VOCs or NOx
	Severe	25 tpy VOCs or NOx
	Extreme	10 tpy VOCs or NOx
	Transport regions not classified as severe or extreme	50 tpy VOCs
Carbon monoxide	Moderate	100 tpy CO
	Serious	50 tpy CO
PM-10	Moderate	100 tpy PM-10
	Serious	70 tpy PM-10

[1] EPA has authority to classify SOx, NO_2, and lead nonattainment areas by seriousness of the nonattainment problem, in order to apply attainment dates and other relevant criteria (CAA Section 172(a)(1)(A). Currently, EPA has no plans to establish a classification scheme for SO2 nonattainment areas (56 FR 13545).

[2] In ozone nonattainment areas, the major source threshold applies to VOC or NOx emissions, but not a sum of those emissions. For example, a source in a severe nonattainment area that emits 20 tpy of VOCs and 20 tpy of NOx is not considered a major source.

Source: *Clean Air Handbook 3rd Edition.* Government Institutes, Inc. 1998.

6.3.3   Air Pollution Permits

Permits can take several forms. These include the two discussed below - an **operating permit (Title V)** and a **preconstruction permit**, also known as a new source permit.

Permit Type: Title V Operating Permit

An operating permit (Title V) will contain all applicable and enforceable control requirements and, like all permits, will have a defined period of effectiveness. An operating permit serves three purposes:

(1)   Provides an inventory of air pollution emission units at sources. This inventory is used by federal, state, and local agencies to plan for either further reductions of air pollution or the maintenance of current air quality.

(2)   Indicates the control requirements to be used to reduce the emissions of pollutants at a facility.

(3)   Identifies how a facility demonstrates compliance.

The Title V operating permit specifies all of the applicable state and federal requirements, including emission limits; and recordkeeping, monitoring, and reporting requirements with which your facility must comply. It also has a defined period of effectiveness. You must obtain a Title V operating permit if you are a:

> *Monitoring, recordkeeping, and reporting requirements for operating permits can be found at 40 CFR 70.6.*

• Major source with the potential to emit 100 or more tpy of any air pollutant in <u>attainment areas</u> (as discussed in Section 6.3.2 *Determining Whether Your Facility Meets Federal Regulations*).

• Major source with the potential to emit ten or more tpy of any one HAP, or 25 tpy of any combination of HAPs in attainment areas (as discussed in Section 6.3.2 *Determining Whether Your Facility Meets Federal Regulations*). Note: a non-major source of HAPs may be still be required to obtain a permit under NESHAP, see below.

• Major source subject to nonattainment provisions where lower thresholds apply depending on an area's severity classification for ozone, carbon monoxide or particulate matter.

• Facility subject to NSPS (40 CFR 60) or NESHAP (40 CFR 61 & 63).

• Facility required to have a pre-construction permit in a nonattainment area or prevention of significant deterioration area.

Although operating permits are issued by the states, EPA is authorized to review and approve the state's permit program, as well as to review and approve each individual permit issued by the state.

Minor sources. Generally, there are two types of minor sources: **natural minor sources** and **synthetic minor sources**.

- **Natural minor sources** are facilities whose potential to emit are below applicable thresholds without any restrictions on operations or enforceable control technology. If your facility is a natural minor source, you are not subject to **major source** federal requirements. However, the state may require you to obtain a federally enforceable state operating permit (FESOP) (54 FR 27274). These state permits may require certain requirements, such as restrictions on production, hours of operation, and recordkeeping and reporting provisions.

- **Synthetic minor sources** are considered to be minor sources after installing restrictions on operations or enforceable control technlogy. A facility may declare itself to be a synthetic minor source if its potential to emit is less than the applicable thresholds and the permitting authority approves this declaration. With this approval, the facility accepts emissions limits, or installs control technology to achieve emissions reductions that allow the permitting authority to consider the facility a minor source.

Note: There are pollutants that are subject to EPA and/or state regulations regardless of a source's size. For example, solvent degreasers (used to clean machinery), are commonly subject to regulations regardless of a source's size.

Compliance Assurance Monitoring. EPA issued its final Compliance Assurance Monitoring (CAM) rule in October 1997 (40 CFR 64) in order to satisfy the requirements for monitoring and compliance certification under the Title V Operating Permits program (40 CFR 70) and the CAAA of 1990. The purpose of CAM is to help owners and operators of facilities to conduct effective monitoring of air pollution control equipment. Under CAM, you must monitor the operation and maintenance of your control equipment in order to evaluate the performance of control devices and report whether or not your facility meets established emission standards. If you find that your control equipment is not working properly, the CAM rule requires you to take action to correct any malfunctions and to report such instances to the appropriate enforcement agency. For more information on CAM, refer to EPA's website at **http://www.epa.gov/ttnuatw1/cam/**.

Permit Type: Pre-construction Permit (also known as a New Source Permit)

A **pre-construction permit** is required before a new major emissions unit(s) can be built in a nonattainment area. A pre-construction permit is often called a **construction permit** or a **permit to install (PTI)**. Section 110 of the CAA regulates construction of major new sources or major modifications of existing sources in **nonattainment** areas through its **New Source Review (NSR) Program**. Your facility is

> *Note: New source review provisions (CAA Section 110),* which are required for state implementation plans, are administered independently from **new source performance standards (NSPS),** which authorize EPA to identify categories of new and modified sources that contribute significantly to air pollution that endangers human health or welfare (CAA Section 111).

subject to NSR permitting requirements if you are either a new major source or an existing major source with significant modifications to equipment (e.g., for process operations) at your facility. States are required to implement NSR provisions in their SIPs (40 CFR 51). **Each state's regulations define when a facility is considered a new source.**

Under this program, major new sources or major modifications of existing sources in nonattainment areas must install control technology that will achieve a standard defined as the **Lowest Achievable Emission Rate (LAER)**. NSR also requires major new or modified sources in nonattainment areas to **offset** their emissions. You can offset new emissions by buying or trading emissions reductions from other sources. Most minor new source review programs do not require offsets, but many require the source to implement the **Best Available Control Technology (BACT)**.

Minor New Source Review. If you are a new source whose emissions are less than the threshold(s) for classification as a major, you may still be subject to minor new source review depending on your state. Because you may be more likely to modify your existing facility rather than build a new one, you should understand the regulatory implications of modifying your plant. **Each state has a federally approved program to regulate minor modifications and minor new sources.**

Prevention of Significant Deterioration. The Prevention of Significant Deterioration (PSD) program applies to facilities in **attainment** areas. Under the PSD program, certain types of stationary sources with the potential to emit more than 100 tons per year of any regulated pollutant or any source that emits more than 250 tons per year of any one pollutant may be required to obtain a PSD permit. The permit must be obtained before construction of a **major** new source or a **major modification of an existing source** takes place. In order to obtain the PSD permit, the owner or operator of the facility must demonstrate that the proposed source will (1) comply with NAAQS and PSD **increments** (listed at 40 CFR 50.21); (2) employ best available control technology for regulated pollutants emitted in significant amounts, and (3) have no adverse impact on other air quality related values. **Note: PSD permitting requirements generally do not affect many food processing facilities; however, check with your state permitting authority.**

> *In order to obtain a PSD permit, your facility must demonstrate that it will employ Best Available Control Technology (BACT). BACT is defined as the maximum degree of emission reduction achievable and takes into account economic, energy, and environmental factors.*

6.4 Risk Management Planning

As required under Section 112(r) of the amended CAA, EPA has promulgated the Risk Management Program Rule. The rule's main goals are to prevent accidental releases of regulated substances and to reduce the severity of those releases that do occur by requiring facilities to develop risk management programs. A facility's risk management program must incorporate three elements: a hazard assessment, a prevention program, and an emergency response program. These programs are to be summarized in a risk management plan (RMP) that will be made available to state and local government agencies and the public. Besides helping facilities prevent accidents, the rule can improve the efficiency of work operations by ensuring that workers are trained in proper procedures and by using preventive maintenance to reduce equipment breakdowns.

Who's Covered. If you have more than a threshold quantity of any of the **regulated substances** in a single process, you are required to comply with the regulation (40 CFR 68). In terms of this regulation, process means "manufacturing, storing, distributing, handling, or using" a regulated substance in any other way. Ammonia, chlorine, and propane are some of the regulated substances commonly used by food processing facilities. **Covered facilities must comply with the rule by June 21, 1999.**

> *EPA has currently established a list of 140 regulated substances covered by these CAA regulations. These substances were published in the Federal Register on January 31, 1994; EPA amended the list by rule, published on December 18, 1997. EPA may amend the list in the future as needed.*

Three levels of compliance. The risk management planning regulation (40 CFR Part 68) defines the activities facilities must undertake to address the risks posed by regulated substances in covered processes. To ensure that individual processes are subject to appropriate requirements that match their size and the risks they may pose, EPA has classified them into 3 categories ("programs"):

> *A risk management program is similar to OSHA's **Process Safety Management (PSM)** program for highly hazardous chemicals (29 CFR 1910.119) that became effective in May 1992. The difference between the programs is the focus. The OSHA regulation is concerned with worker safety, while EPA's CAA regulation is concerned with the safety of the environment and community. For more information about inventory and reporting requirements for OSHA hazardous chemicals, see Section 7.4 Hazard Chemical Inventory and Reporting.*

- **Program 1** requirements apply to processes for which a worst-case release, as evaluated in the hazard assessment, would not affect the public. These are processes that have **not** had an accidental release that caused serious offsite consequences.

- **Program 2** requirements apply to less complex operations that do **not** involve chemical processing.

- **Program 3** requirements apply to higher risk, complex chemical processing operations and to processes already subject to the **OSHA Process Safety Management Standard (29 CFR 1910.119)**.

Risk Management Planning. If your facility has more than a threshold quantity of any of the 140 regulated substances in a single process, you are required to develop a risk management program and to summarize your program in a risk management plan by June 21, 1999. If you are a facility with processes in Program 1, you must carry out the following elements of risk management planning:

- An offsite consequence analysis that evaluates specific potential release scenarios, including worst-case and alternative scenarios.

- A five-year history of certain accidental releases of regulated substances from covered processes.

- A risk management plan (RMP), revised at least once every five years, that summarizes and documents these activities for all covered processes.

Facilities with processes in Programs 2 and 3 must also address each of the following elements:

- An integrated prevention program to manage risk. The prevention program will include identification of hazards, written operating procedures, training, maintenance, and accident investigation.

•• An emergency response program.

- An overall management system to supervise the implementation of these program elements.

Risk Management Plan. If you do not already have a risk management plan, you should develop one as soon as possible. Your plan may include some or all of the following elements:

- Documentation of process safety information
- Process hazard analysis information
- Documentation of operating procedures
- Training program information
- Pre-startup review information
- Maintenance program information
- Management of Change program information
- Accident history
- Emergency response program information
- Worst-case and alternative release scenarios
•• Other elements

The plan you submit to EPA will summarize your program and will have to be made available to the public. (Note: EPA's deadline for determining whether facilities must submit their RMPs to EPA Headquarters or to the regional offices is June 21, 1999.) Once your plan is submitted, it will be reviewed for accuracy and completeness. A site visit also may be conducted at your facility by either EPA, state, or local officials to determine whether your plan accurately reflects your risk management program in operation.

Industry-specific guidance. To make compliance easier for small businesses, EPA has worked with trade associations and other industry groups to develop a series of industry-specific brochures that will assist businesses in creating their risk management programs. Of these, the brochures that may be applicable to the food processing industry include those for water treatment facilities (i.e., publicly owned treatment works), propane users, and operators of ammonia refrigeration systems. These brochures can be accessed at EPA's Chemical Accident Prevention and Risk Management Planning website at **http://www.epa.gov/ceppo/**.

> **Excerpt from EPA's RMP Brochure for Operators of Ammonia Refrigeration Systems**
> Under the Risk Management Program Rule, some operators of ammonia refrigeration systems will have to implement a risk management program and file a risk management plan (RMP) with EPA by June 21, 1999. If you store or use a total of more than 10,000 lbs of ammonia at your facility in one or more interconnected tanks, receiver vessels, or pipelines, you are likely to be subject to this rule. If you operate two refrigeration systems with adjacent equipment, consider the total quantity of ammonia in both systems when determining is this rule applies to you. For more information, access this brochure at **http://www.epa.gov/ceppo/** or see Section 6.5.2 Air Conditioners/Refrigeration Service and Disposal: Ammonia and CFCs.

Model Risk Management Programs. EPA has been working with industry groups to develop model risk management programs. One of these is for ammonia refrigeration systems. To review this model program, refer to EPA's Chemical Accident Prevention and Risk Management Planning website at **http://www.epa.gov/swercepp/acc-pre.htm#Model Plans/**.

Communicating RMP Requirements. The Food Industry Environmental Council (FIEC), a coalition of more than 50 food processors and trade associations, has developed materials to assist food processors in communicating with the public about risk management programs. These communication materials include the following:

- •• "Backgrounders" on ammonia, chlorine, and propane;
- •• A computer disk with the shell of a tri-fold brochure and filler language;
- •• Communication guidelines;
- •• A question and answer document; and
- •• A resource and reference document.

The communication packages are available from your food trade association.

For more information about risk management planning requirements, visit EPA's Chemical Emergency Preparedness and Prevention Office's webpage at **http://www.epa.gov/ceppo/**) or refer to Section 9.2 *Emergency Planning and Reporting Requirements*. You also may obtain copies of the rule and a wide variety of technical assistance materials, as well as answers to your specific questions, from **EPA's RCRA/UST, Superfund and EPCRA hotline at 1-800-424-9346 or 703-412-9810.**

6.5 Air Compliance Issues for Selected Operations

6.5.1 Boilers or Steam Generating Units

Most food processing facilities have industrial boilers or hot water heaters for generating steam or hot water for processing, cooking, or sanitation. Industrial boilers tend to be smaller in size, subject to more and greater load swings, operated at a lower capacity factor, and capable of utilizing multiple fuels. In addition, they often are the only supplier to their site and must be highly reliable. Coal, fuel oil, and natural gas are the major fossil fuels used by boilers. The combustion of these fossil fuels produces primarily sulfur oxides (SOx), nitrogen oxides (NOx) and particulate emissions nationwide, with minor amounts of VOCs and carbon monoxide.

> *If your facility stores fuel oil onsite, you must comply with the Oil Pollution Act's regulations. For more information, see Section 4.6 The Oil Pollution Act Regulation.*

If your facility has any of the following types of boilers, then you must comply with federal emission limits for NOx, SO_2, and particulates:

(1) A fossil fuel-fired or fossil fuel and wood residue-fired steam generator which has a heat input rate of more than 250 million Btu and was constructed after August 17, 1971 (40 CFR 60 Subpart D);

(2) An industrial-commercial-institutional (ICI) steam generator which has a heat input rate of more than 100 million Btu and was constructed, modified, or reconstructed after June 19, 1984 (40 CFR 60 Subpart Db); or

(3) A small ICI generator which has a heat input capacity ranging from 10 million Btu to 100 million Btu per hour or less and was constructed, modified, or reconstructed after June 9, 1989 (40 CFR 60 Subpart Dc).

As stated above, NOx emissions are common type of emissions from boilers and these emissions must meet federal limits. Table 6-2 *Federal Emission Standards for NOx* summarizes the federal NOx emission limits for the first two types of boilers listed above. Refer to 40 CFR 60, Subpart Dc for information on the third type of boiler listed above. Similar emission limits for SO_2 and particulates can be found in 40 CFR 60 Subparts D, Db, and Dc.

State Standards. In addition to the federal emission limits for NOx, SO_2, and PM, state and local governments may have additional or more stringent emission limits. State emission standards for boilers vary depending on the **attainment** status of the geographical region as well as other factors (see Section 6.2 *What is the Clean Air Act?*). For example, while some states such as South Dakota defer to federal regulations when setting emission limits for steam generators, other states implement more stringent regulations. Also states, such as Pennsylvania and Massachusetts, have implemented NOx emission trading programs that may affect different types and sizes of boilers within their states. **Contact your state regulatory agency for more information on state emission limits.**

Activities Related to Emission Limits. EPA is leading or participating in several major activities related to emission limits for NOx and VOCs, which will affect the regulation of steam generating units. Chief among these are the Ozone Transport Assessment Group (OTAG) and the Industrial Combustion Coordinated Rulemaking (ICCR) that are described below:

- **Ozone Transport Assessment Group (OTAG)**. To assist with compliance with NAAQS, OTAG is identifying and recommending to EPA cost-effective control strategies for NOx and VOCs. OTAG, which is a partnership between EPA, the Environmental Council of the States (ECOS), and various industry and environmental groups, prepared the *Assessment of Control Technologies for Reducing Nitrogen Oxide Emissions From Non-Utility Point Sources and Major Source Areas - Appendix C*. This report provides an overview of NOx control technologies available for non-utility, fossil-fuel fired boilers and can be reviewed at EPA's website at **http://www.epa.gov/ttnotag1/finalrpt/ chp5/appc.htm/**.

- **Industrial Combustion Coordinated Rulemaking (ICCR)**. EPA is planning an ICCR for ICI combustion sources (e.g., boilers, process heaters, waste incinerators). EPA will develop recommendations for federal air emission regulations that address the various combustion source categories and pollutants. These regulations will be developed under Sections 111 (NSPS), 112 (NESHAP), and 129 (solid waste combustion) of the CAA. Seven categories of ICI combustion sources are listed for regulatory development as follows:

  – Industrial boilers (Sections 111 and 112);
  – Commercial-institutional boilers (Sections 111 and 112);
  – Process heaters (Sections 111 and 112);
  – Industrial-commercial solid waste incinerators (Sections 111 and 129);
  – Other solid waste incinerators (Sections 111 and 129);
  – Stationary combustion turbines (Sections 111 and 112); and
  – Stationary internal combustion engines (Sections 111 and 112).

  The overall goal of the ICCR is to reduce the potential for conflicting or duplicative regulations for the various combustion source categories, rather than regulating each source category individually. This approach will facilitate consistency and produce greater environmental benefits at lower cost. For more information on the status of ICCR, refer to EPA's website at **http://www.epa.gov/ttn/iccr/**.

Table 6-2. Federal Emission Standards for NOx (Emission limits for SO$_2$ and PM can be found in 40 CFR 60, Subparts D, Db, and Dc.)

Fossil-Fuel-Fired Steam Generators with a heat input rate > 250 million Btu per hour constructed or modified after August 17, 1971 (40 CFR 60 Subpart D)	
Fuel Type	**Emission Limit**
Gaseous fossil fuel	0.20 lb/Mbtu
Liquid fossil fuel[1]	0.30 lb/Mbtu
Solid fossil fuel (mixed with or without wood residue)[1]	0.70 lb/Mbtu
Lignite or lignite and wood residue[1]	0.60 lb/Mbtu

For Industrial-Commercial-Institutional Steam Generating Units with a heat input rate > 100 million Btu per hour constructed, modified, or reconstructed after June 19, 1984 (40 CFR 60 Subpart Db)	
Fuel Type	**Emission Limit**
Natural Gas and distillate oil 1) Low heat release rate 2) High heat release rate	 0.10 lb/Mbtu 0.20 lb/Mbtu
Residual Oil 1) Low heat release rate 2) High heat release rate	 0.30 lb/Mbtu 0.40 lb/Mbtu
Coal 1) Mass-feed stoker 2) Spreader stoker and fluidized bed combustion 3) Pulverized coal 4) Lignite (except lignite mined in ND, SD, or MT) 5) Lignite mined in ND, SD, or MT and combusted in a slag tap furnace 6) Coal-derived synthetic fuels	 0.50 lb/Mbtu 0.60 lb/Mbtu 0.70 lb/Mbtu 0.60 lb/Mbtu 0.80 lb/Mbtu 0.50 lb/Mbtu
Duct burner in system: 1) Natural gas and distillate oil 2) Residual oil	 0.20 lb/Mbtu 0.40 lb/Mbtu
Mixtures of coal, oil, or natural gas.	Refer to formula defined in 40 CFR 60 Subpart Db; Section 60.44b.
Coal or oil, or a mixture with other fuels.	Refer to formula listed in 40 CFR 60 Subpart Db; Section 60.44b.
Natural gas mixed with wood, municipal-type solid waste, or other solid fuel (except coal).	0.30 lb/Mbtu[2]
Coal, oil, or natural gas mixed with by-product/waste.	Refer to formula defined in 40 CFR 60 Subpart Db; Section 40.44b.[3]

[1] See 40 CFR 60, Subpart Db Section 60.44 for specific exceptions to these emission limits.
[2] Does not apply if facility has an annual capacity factor of 10% or less for natural gas and is subject to federally enforceable requirement limiting operations to an annual capacity factor of 10% or less.
[3] Does not apply if facility has an annual capacity factor of 10% or less for coal, oil, and natural gas and subject to federally enforceable requirement limiting operations to an annual capacity factor of 10% or less.

6.5.2 Air Conditioners/Refrigeration Service and Disposal: Ammonia and CFCs

Ammonia

Most food processing facilities use closed loop ammonia refrigeration systems for heat exchange. Ammonia is handled as a gas and must be added to refrigeration systems to replace amounts lost through leaks or because of losses when purging a section of the system for maintenance. Because ammonia is not a listed air pollutant or classified as one of the 188 hazardous air pollutants, a Title V operating permit for ammonia emissions is not likely to be required. However, it is possible that ammonia will be subject to state permitting requirements.

Ammonia refrigeration systems are subject to Section 112(r) of the amended CAA, which mandates EPA to publish rules and guidance for chemical accident prevention. Ammonia is a volatile chemical and will be released to air through system filling, relief vents, and leaks in valves and fittings. All ammonia lost through these means should be reported as fugitive emissions in a Toxic Release Inventory (TRI) report (40 CFR 372), if the total is over the threshold amount. See Section 7.5 *Toxic Chemical Release Reporting - Section 313* for more information about TRI reporting.

On January 31, 1994, EPA promulgated a final list of 140 regulated substances and threshold quantities, which are identified under Section 112(r). According to the final list, ammonia is a regulated substance if it is at a **concentration of at least 20 percent** and exceeds the established threshold quantity of 20,000 lbs (40 CFR 68). Therefore, if your facility has a **process** that uses a 20 percent ammonia solution which exceeds the threshold quantity established by EPA, you must develop and implement a risk management plan (RMP) for that process. See Section 6.4 *Risk Management Planning* for more information. For a comparison of these requirements to similar requirements under other EPA statutes, see Section 9.2 *Emergency Planning and Reporting Requirements*.

Chlorofluorocarbons (CFCs)

Your food processing facility may be subject to requirements of the stratospheric ozone protection program if you have motor vehicle air conditioners, certain appliances (air conditioners, refrigerators, and freezers), and industrial process refrigeration units that use CFCs and other class I and class II substances (see box).

The CAA provides a framework for the regulation of ozone-depleting substances such as CFCs to protect the stratospheric ozone layer. EPA's stratospheric ozone regulation does the following:

- Bans the use of certain ozone-depleting substances in non-essential products;

- Requires labels for products containing or manufactured with regulated ozone-depleting substances;

- Bans the production of many of these substances (see 40 CFR 82).

EPA has established requirements for servicing and disposal of air-conditioning and refrigeration equipment that contains regulated ozone-depleting refrigerants. These requirements described briefly below are intended to minimize the release of such refrigerants to the atmosphere. If you own/operate appliances containing ozone-depleting refrigerants, you must do the following:

> *Ozone-depleting chemicals to be regulated have been divided into two classes based on their ozone depletion potential:*
>
> ***Class I*** *includes specified CFCs, halons, methyl chloroform, carbon tetrachloride, methyl bromide, and HBFCs. Production of these chemicals were phased out in 1996, except for methyl bromide, production of which will be banned in 2001.*
>
> *The principal substances included in **Class II** are hydrochlorofluorocarbons (HCFCs). Some HCFCs will be phased out either partially (HCFC-22, HCFC-1426) or entirely (HCFC-1416) beginning in 2003. The HCFCs with the most severe ozone-depleting effects will be phased out first. **Note that the phase out is for production and importation, not use.** Thus HCFCs can be used as refrigerants after 2020; however, they may not be available.*

- When opening any appliance containing refrigerants for maintenance, service, repair, or disposal, you must have at least one piece of certified, self-contained recovery equipment available at your facility.

- Notify EPA that such equipment is available at your facility. This equipment must be operated to certain specified standards that minimize atmospheric release of refrigerants.

- If your appliances contain 50 or more pounds of refrigerant, you must repair leaks in a timely manner. You must maintain records documenting the date and type of all servicing performed on the appliance, as well as the quantity of refrigerant added.

- If you are an appliance owner/operator who adds the refrigerant, you must maintain records of refrigerant purchased and added.

- If you use technicians to service and maintain refrigerant-containing appliances, they must be certified by an approved technician certification program.

- If you employ such technicians, you must maintain records demonstrating compliance with the certification requirement (see 40 CFR 82).

In addition to federal regulations, many state and local governments have enacted legislation and ordinances limiting the production and use of ozone-depleting substances. Contact your state permitting authority to find out about all requirements that apply to you. For information about EPA's Stratospheric Ozone Protection Program [including EPA's *Significant New Alternative Policy* (SNAP) program], call the Stratospheric Ozone Hotline at 1-800-296-1996 or visit EPA's website at **http://www.epa.gov/ozone/**.

6.5.3   Building Renovation/Demolition: Asbestos

If you are renovating or demolishing a structure on your property, you have the potential to release asbestos fibers that can create serious worker health and safety problems. Asbestos is an insulation material widely used in the past where fire retardation was required or desirable. Applications for asbestos include, but are not limited to, floor tiles, ceiling tiles, siding, and thermal system insulation. Renovations or demolition activities involving asbestos-containing materials are regulated by the CAA's NESHAPs (see CAA Section 112; 40 CFR 61, Subpart M). Although considered a serious health hazard, asbestos is not regulated as a RCRA hazardous waste.

Asbestos fibers have been linked to serious adverse health effects from the inhalation of airborne asbestos fibers. However, if asbestos is present in your facility, it does not mean that your employees are in danger. As long as the material containing the asbestos remains in good condition and is not disturbed, exposure to asbestos fibers is unlikely. The threat of exposure arises when asbestos-containing

> *The chances for human exposure to asbestos are highest during maintenance work or building demolition.*

materials are disturbed through repair, renovation, demolition, or natural disturbances, and asbestos fibers potentially are released. Government regulations now are requiring that asbestos be phased out of production and use.

If you are planning any renovation or demolition activities, you should assume that most old building construction materials contain asbestos. Typical asbestos-containing materials include pipe and duct insulation, fireproofing, roofing materials, floor tile, and transite pipe and sheet goods. Many other building materials, such as ceiling tiles, wall board, plasters, and fire doors, may also contain asbestos.

If you are planning any renovation or demolition activity at your facility, you should:

- Contact your regional environmental agency (in some cities or counties, this may be the health department) before renovating or demolishing a building or structure, regardless of whether asbestos-containing material is present or only suspected.

- Remove asbestos-containing materials using only qualified personnel in accordance with all applicable local, state, and federal laws. This material must be removed *prior* to any demolition or renovation activity. It is recommended that you review your contractor's employee training records and licenses.

- Use special handling procedures for asbestos disposal such as asbestos certified contractors for assessment and demolition of pre-1980 buildings, as well as posting of signs at disposal facilities.

For demolition activities, many states have a formal notification process before demolition may begin. For example, Ohio requires at least a 10 day notice before any demolition or construction activity begins. Other requirements may include inspection by a licensed building inspector before construction or demolition may begin. **Check with your state and local authorities to determine whether additional asbestos requirements apply to you.**

6.5.4   Odor Emissions

The combination of a broad regulatory framework, increased sensitivity and demand of the general public for a clean and pleasant environment, and reduced land areas available for isolation of industrial operations from the public areas have forced all types of industries including the food processing industry to control odor emissions. Organic and inorganic compounds emitted from various food processing operations may become nuisances in your community when they carry objectionable odors as perceived by the general public. Though there are no federal regulations for odor emissions, you should be aware that there may be state and local regulations.

There are two basic principles for controlling odors at a food processing plant:

- • • Reduction of odors at the generation sources

- • Removal of odors from collection air-streams before the odors are discharged into the atmosphere.

Odors generated from food processing plants usually are a mixture of various organic and inorganic compounds in low concentrations. Most of these compounds are reduced carbon, nitrogen and/or sulfur compounds. Typical odorous compounds encountered in food processing operations include aldehydes, ketones, alcohols, acids,

> *For more information on odor emissions, refer to **Odor Control and Wastewater Treatment,** published by Water Environment Federation and American Society of Civil Engineers (1995).*

ammonia, amines, sulfides, mercaptans, and hydrogen sulfide. In some cases, the odors also may be caused by VOCs (e.g., VOCs from drying and roasting activities) which are less biodegradable. The physical and chemical characteristics of specific odors are affected largely by the types of odor sources. Effective, application-specific air cleaning technologies are needed to help food processors make their operations environmentally friendly.

SECTION 7 CONTENTS

7. HOW DO I COMPLY WITH THE EMERGENCY PLANNING AND COMMUNITY RIGHT-TO-KNOW ACT REGULATIONS?

7.1 Introduction

This section presents an overview of the Emergency Planning and Community Right-to-Know Act (EPCRA) planning, reporting, and notification requirements for your food processing facility. Because of concern over the Bhopal tragedy of 1984 and many state and community right-to-know laws, Congress passed Title III of the Superfund Amendments and Reauthorization Act (SARA Title III) in 1986. Title III of SARA, also known as EPCRA, establishes requirements for federal, state, and local governments and industry regarding emergency planning and "community right-to-know" reporting on hazardous and toxic chemicals. To this end, it requires industry to report detailed information concerning the use, generation, release, and other waste management activities of hazardous and toxic materials.

EPCRA is unique compared to other environmental statutes because it does not establish release limitations, standards of practice, or standards of operation for industry. The purpose of EPCRA is to:

- Encourage and support industry's emergency planning for response to chemical accidents (in coordination with state and local governments) through emergency planning and emergency notification; and

- Provide local governments and the public with information about possible chemical hazards in their communities by requiring facilities to (1) report to their State Emergency Response Commissions (SERCs), Local Emergency Planning Committees (LEPCs), and local fire departments their hazardous chemical inventory, and (2) report to federal and state authorities their toxic chemical releases and other waste management practices.

> *Under the emergency planning requirements of EPCRA, each state governor must appoint a SERC. Each SERC in turn appoints LEPCs. For addresses for these groups, see Appendix B of this guide. For more information, access* ***http:// www.epa.gov/ceppo/***.

Your facility may be subject to emergency planning, reporting, notification, and response requirements under EPCRA including:

- •• Emergency planning (Sections 301-303)
- •• Emergency release notification (EPCRA Section 304 and Comprehensive Environmental Response, Compensation, and Liability Act (CERCLA) Section 103)

•• Hazardous chemical inventory and reporting (MSDS and Tier reporting) (Sections 311 and 312)
•• Toxic chemical release reporting (Section 313).

The particular substances subject to these EPCRA requirements are defined under 3 statutes; EPCRA, CERCLA, and Occupational Safety and Health Administration (OSHA). They are identified using various terms as shown in Table 7-1 below. Also, the types of substances subject to each of the EPCRA requirements vary. Refer to the sections indicated in Table 7-1 for more information.

Table 7-1. Guide to Substances Subject to EPCRA

Section of the Law	Types of Substances Regulated	CFR Citation
Emergency Planning EPCRA Sections 301-303	EPCRA extremely hazardous substances	40 CFR 355
Emergency Release Notification EPCRA Section 304/CERCLA Section 103	EPCRA extremely hazardous substances	40 CFR 355
	CERCLA hazardous substances	40 CFR 302
Hazardous Chemical Inventory and Reporting EPCRA Sections 311 and 312	EPCRA extremely hazardous substances	40 CFR 370
	OSHA hazardous chemicals	29 CFR 1910
Toxic Chemical Release Reporting EPCRA Section 313	Toxic chemicals	40 CFR 372

Keep in mind the following distinctions among the EPCRA sections; EPCRA Sections 301-303, 311, and 312 focus on chemicals **present at** your facility, whereas EPCRA Section 313 focuses on chemical **manufactured, processed, or otherwise used**. EPCRA Section 304 focuses on emergency notification of a **release** of specific substances.

It is important to note that if you eliminate EPCRA chemicals from your operations through pollution prevention (P2) activities, you also will eliminate the associated planning and notification requirements. P2 is an excellent opportunity to decrease your facility's regulatory burden.

Section 7.2 summarizes the principal planning and reporting requirements for EPCRA Sections 301-303. Section 7.3 presents the emergency notification and release reporting requirements under EPCRA Section 304 and CERCLA Section 103. Section 7.4 presents the hazardous chemical inventory and reporting requirements under EPCRA Sections 311 and 312.

Finally, Section 7.5 presents an overview of the EPCRA Section 313 reporting requirements and estimation of releases and other waste management quantities. EPA's Office of Pollution Prevention and Toxics (OPPT) has developed substantive guidance for food processors on compliance with EPCRA 313, entitled, *EPCRA Section 313 Reporting*

> **New Guidance**: *For more information on EPCRA 313 requirements, see EPCRA Section 313 Reporting Guidance for Food Processors (EPA 745-R-98-011, September 1998).*

Guidance for Food Processors (EPA 745-R-98-011, September 1998). The text in Section 7.5 is excerpted from OPPT's guidance for this industry sector. Please refer to the OPPT guidance document for additional information.

7.2 Emergency Planning

The emergency planning sections (Sections 301-303) of EPCRA are designed to develop state and local governments' emergency response and preparedness capabilities through better coordination and planning, especially with the local community.

Under Section 302 of EPCRA, if your food processing facility, no matter how small, has any of the extremely hazardous substances (EHSs) listed in 40 CFR 355 in amounts equal to or in excess of certain minimum amounts (called threshold planning quantities [TPQs]), you must participate in emergency planning activities. EHSs typically found at food processing facilities include ammonia (for refrigeration), chlorine (for disinfection), and nitric and sulfuric acids (for cleaning). The threshold planning and spill/release *reportable quantities* (see side box) for these chemicals are listed below.

> A ***threshold planning quantity*** *(TPQ) is the amount of an EHS, in pounds, at a facility that triggers a reporting requirement. EHSs and their TPQs are listed in 40 CFR 355.*
>
> A ***reportable quantity*** *(RQ) is the amount of an EHS or CERCLA hazardous substance released into the environment within a 24-hour period. RQs for EHSs are found in 40 CFR 355, Appendices A and B. RQs for CERCLA hazardous substances are found in 40 CFR 302, Table 302.4. The RQ for any other substance is one pound.*

Extremely Hazardous Substances	Threshold Planning Quantity (lbs)	Reportable Quantity (lbs)
Ammonia	500	100
Chlorine	100	10
Nitric Acid	1,000	1,000
Sulfuric Acid	1,000	1,000

If your facility has any of the EHSs onsite in quantities equal to or greater than the TPQs, you must notify the SERC and LEPC within 60 days after the EHSs are present in these quantities. For more information on EPCRA Section 302 reporting requirements, contact the RCRA/UST, Superfund and EPCRA Hotline at 1-800-424-9346 or 703-412-

> ***Blended Chemicals:*** *When calculating amounts of blended chemicals, it is important to note that only the specific portion of the blend which contains the EHS is counted, not the whole blend. For example, 100 lbs of a 20 percent chlorine compound counts as 20 lbs, not 100 lbs, of chlorine.*

9810, or access EPA's Chemical Emergency Preparedness and Prevention Office homepage at **http://www.epa.gov/ceppo/**.

7.3 Emergency Release Notification

The emergency release notification requirements set out in EPCRA and CERCLA enable federal, state, and local authorities to effectively prepare for and respond to chemical accidents.

The release notification requirements differ slightly between the two laws, but the requirements are interrelated as explained below. Releases of both EPCRA EHSs and CERCLA hazardous substances are reportable under EPCRA Section 304, whereas only releases of CERCLA hazardous substances are reportable under CERCLA. Another difference between the statutes is EPCRA requires that Section 304 release notifications be provided to SERCs and LEPCs, whereas CERCLA requires that Section 103 release notifications be provided to the National Response Center (NRC).

> **What is the NRC?** *The primary function of the National Response Center (NRC) is to serve as the sole national (federal) point of contact for reporting all oil, chemical, and other discharges into the environment anywhere in the U.S. and its territories. For more information on the NRC, access* **http://www.epa.gov/oilspill/**.

Releases and Reportable Quantities. The first step in determining if release notification requirements are triggered is assessing whether or not a release has occurred.   Under EPCRA, a release is as any spilling, leaking, pumping, pouring, emitting, emptying, discharging, injecting, escaping, leaching, dumping, or disposing into the environment, including abandonment or discharging of barrels, containers, and other closed receptacles containing any hazardous substance, pollutant, or contaminant. EPCRA's definition includes releases of both EPCRA EHSs and CERCLA hazardous substances (40 CFR 355.20), and EPCRA Section 304 carries an additional requirement that a facility must produce, use, or store the substance in order to have a **reportable release**. The list of EHSs can be found at 40 CFR 355, Appendices A and B. The term **hazardous substance** is defined in CERCLA 101(14), and these substances are listed in 40 CFR 302, Table 302.4.

In order for a release of a EHS or CERCLA hazardous substance to be reportable, a certain amount must be released into the environment within a 24-hour period. This amount, called the **reportable quantity (RQ)**, triggers emergency release notification requirements. For each CERCLA hazardous substance and EHS identified, EPA has designated a reportable quantity (RQ) of 1, 10, 100, 1,000, or 5,000 pounds. Reportable quantities are listed in 40 CFR 355, Appendices A and B.

Notification. In order to ensure proper and immediate responses to potential chemical hazardous, EPCRA Section 304 requires facilities to **notify SERCS and LEPCs** of releases of EPCRA EHSs and CERCLA hazardous substances when the release equals or exceeds the RQ (EPCRA 304(a)). To trigger EPCRA Section 304 notification, there must be:

- •• A facility at which a hazardous chemical is produced, used, or stored; **AND** (all of the following)
- • A release

- Of an EHS or CERCLA hazardous substance
- Into the environment
- With a potential to affect human health and the environment offsite
- That equals or exceeds a reportable quantity
- Within a 24-hour period.

The LEPCs and SERCs will coordinate response activity to your spill or accident, and prevent harmful effects to the public. These agencies also may provide instructions to you regarding appropriate response procedures.

Additionally, when there is a release of a CERCLA hazardous substance in an amount equal to or in excess of the RQ for that substance (CERCLA 103(a), 40 CFR 302.6), CERCLA requires the person in charge of a vessel or facility to immediately notify the **National Response Center at 1-800-424-8802.** There are six specific conditions that must be met to trigger the CERCLA requirement for notifying the National Response Center. There must be a:

- Release
- Of a CERCLA hazardous substance
- Into the environment
- From a vessel or facility
- That equals or exceeds a reportable quantity
- Within a 24-hour period.

Releases That Are Not Reportable. There are several types of releases that are excluded from the requirements of both EPCRA and CERCLA release notification. These releases were excluded originally under CERCLA Section 101(22) because they are covered by other regulatory programs. The regulations found at 40 CFR 355.40(a)(2)(v) extend these statutory exclusions under CERCLA to the release reporting requirements under EPCRA. Examples of these instances are included here for your reference (see box).

When No Notification Is Required (40 CFR 355.40):

1. Releases which result in exposure to persons solely within the boundaries of the facility;

2. Federally permitted releases are not reportable [CERCLA Sections 103(a)and (b) and EPCRA Section 304(a)(2)(A)];

3. Releases that are continuous and stable in quantity and rate (as defined in 40 CFR 302.8(b));

4. Application of pesticide products registered under the Federal Insecticide, Fungicide, and Rodenticide Act (CERCLA Section 103(e));

5. Releases not meeting the definition of release under CERCLA Section 101(22); and

6. Any radionuclide release which occurs naturally in soil.

It is recommended that you make a notification if there is any doubt of applicability because serious fines could result if you are supposed to notify and do not.

Being familiar with your responsibilities for when to report (and when not to report) will help you in responding quickly when a release does occur. When you are required to report, you must complete the initial notification and follow-up actions as discussed below.

Initial Notification. It is very important to know which agency(s) to notify and to do so as soon as practical for any reportable spill. Initial notifications can be made by telephone, radio, or in person. Under EPCRA, initial notification is required **immediately** (see box) upon discovering a spill. Thus the person making the report must use good judgement in determining how much time to spend in collecting information prior to making the notification. This information should include:

> *Although the term "immediately" is not further defined in the regulations, EPA generally defines immediate notification of LEPCs, SERCs, and the National Response Center as within one hour of discovery of a reportable spill or release.*

- •• Chemical name/identity of material(s) released

- • Whether the material(s) is an EPCRA extremely hazardous substance (listed in 40 CFR 355, Appendices A and B) or a CERCLA hazardous substance (listed in 40 CFR 302.4)

- •• Estimate of the quantity of any material released

- •• Time and duration of the release

- •• Whether the release was to the air, water, and/or land

- • Any known or anticipated acute or chronic health risks associated with the emergency, and where necessary, advice regarding medical attention necessary for exposed individuals

- •• Proper precautions, such as evacuation or sheltering in place

- •• Name and telephone number of the person(s) to be contacted for further information.

Follow-up Actions for a Spill or Release. After the initial communication is established with the appropriate agencies, your facility must provide a written follow-up emergency notice, as soon as practicable after the release. The follow-up notice or notices must update information provided in the initial notice and provide information on actual response actions taken, health risks associated with the release, and advice regarding medical attention necessary for exposed individuals.

Your state also may have requirements for notifications and emergency response actions. To identify the appropriate state agencies, call the RCRA/UST, Superfund and EPCRA Hotline at 1-800-424-9346 or 703-412-9810.

7.4 Hazardous Chemical Inventory And Reporting

Moving from the requirements for releases of EPCRA EHSs and CERCLA hazardous substances discussed above, this section addresses the requirements for having EPCRA EHSs and OSHA hazardous chemicals stored on your property.

> *Review this section carefully.* There have been several EPA cases against food processors for failure to comply with EPCRA Section 311 and 312 requirements.

The hazardous chemical inventory and reporting provisions outlined in EPCRA Sections 311 and 312 require you to inventory the hazardous chemicals present onsite at your facility in amounts equal to or in excess of TPQs. This inventory must contain each hazardous chemical's identity, physical and health hazards, and location. There are two reporting mechanisms in the hazardous chemical inventory program:

- A **one-time notification** of the presence of hazardous chemicals onsite in excess of threshold levels (EPCRA Section 311); and

- An **annual notification** detailing the locations and hazards associated with the hazardous chemicals found on facility grounds (EPCRA Section 312).

If your facility meets the applicability criteria described below, you are required to submit these reports to the SERC, LEPC, and local fire department.

Applicability. To be subject to reporting under EPCRA Sections 311 and 312, your facility must meet the applicability criteria (40 CFR 370.20). Applicability is two-fold.

(1) First, your facility must be regulated by the OSHA's Hazardous Communication Standard (HCS).

(2) Second, your facility must exceed EPA-established thresholds for hazardous chemicals onsite at any one time.

OSHA's HCS requires facilities to procure or prepare material safety data sheets (MSDSs) for the hazardous chemicals found at the facility (29 CFR 1910.1200). In general, the chemicals regulated by OSHA's HCS pose a hazard to workers using the substances. **Any facility that is required by OSHA to prepare or have available an MSDS for a hazardous chemical is subject to EPCRA Sections 311 and 312 if the chemical is present onsite at any one time in excess of threshold levels.** There is no list of hazardous chemicals subject to reporting. The key to determining whether or not a chemical is considered hazardous is the requirement to have an MSDS.

Threshold levels. The threshold level varies depending on how the chemical is classified.

- The reporting threshold for hazardous chemicals that are EPCRA extremely hazardous substances (EHSs) is 500 pounds or that chemical's threshold planning quantity (TPQ), whichever is lower. EHSs are listed in 40 CFR 355, Appendix A and B.

- •• The reporting threshold for hazardous chemicals that are **not** EHSs is 10,000 pounds.

Exemptions. Although OSHA requires MSDSs for a large number of chemicals, there are a number of exemptions to the OSHA requirement to maintain MSDSs, **consequently exempting them from EPCRA Sections 311 and 312 reporting**. These are listed in 29 CFR 1910.1200(b)(6).

In addition, SARA Title III Section 311(e) lists five exemptions from the definition of hazardous chemical for purposes of compliance with SARA Title III Sections 311 and 312 (40 CFR 370.2). These exemptions cover chemicals that are either regulated under other programs, do not present a hazard during normal use, are chemicals that the community is already aware of, or are under the control of trained personnel. The exemptions cover:

- •• Food and Drug Administration (FDA) regulated substances (e.g., any food, food additive, color additive, drug, or cosmetic regulated by FDA).
- •• Solid manufactured items.
- •• Substances packaged as consumer products.
- •• Medical and research lab materials.
- •• Substances used in agricultural operations.

It is important to remember that these exemptions apply to specific chemicals within the scope of the exemption only, **not** to all hazardous chemicals at a particular facility.

Section 311 MSDS and Hazardous Chemical Inventory Reporting. Under Section 311 of EPCRA, you must submit a **one-time notification** identifying the hazardous chemicals (including EPCRA extremely hazardous substances and OSHA hazardous chemicals) present at your facility in amounts equal to or in excess of threshold levels to the SERC, LEPC, and local fire department (40 CFR 370.21).

To meet the reporting requirement, your facility must submit the following information for each EPCRA EHS and OSHA hazardous chemical onsite in amounts that equal or exceed the threshold levels, either:

- •• An MSDS (or copies of MSDSs); **or**

- •• A list of the EPCRA EHSs and OSHA hazardous chemicals grouped by hazard category. Hazard categories include immediate health hazard, delayed health hazard, fire hazard, sudden release of pressure hazard, or reactive hazard. The list must include the hazardous chemical name or common name and any hazardous component of each hazardous chemical.

> ***MSDSs:*** *Contact your vendor(s) to obtain MSDSs for chemicals onsite.*

The information needed for compiling this list can be obtained by examining the MSDS for each chemical. Again, the MSDSs or list of hazardous chemicals is a one-time submission and there is no form required by EPA. (You should check with your SERC and LEPC to see if they have a required form.)

If, after initial reporting, your facility finds that it has a hazardous chemical that is newly covered in amounts equal to or excess of the threshold level or there has been significant new information on an already reported chemical, you must update the information reported under Section 311. You must supply this supplemental information within 3 months after discovery of significant new information (40 CFR 370.21(c)).

Section 312 Tier Reporting. Under Section 312 of EPCRA, your facility must meet an **annual reporting requirement** for OSHA hazardous chemicals and EPCRA EHSs in amounts equal to or in excess of threshold levels. If equaling or exceeding the threshold levels at any time in the preceding year, you must submit to the SERC, LEPC, and local fire department an "Emergency and Hazardous Chemical Inventory Form." This form must be submitted by March 1 and covers the previous calendar year.

The reporting thresholds are the same as for submission of MSDSs under EPCRA Section 311: 500 pounds or the TPQ (whichever is lower) for EPCRA EHSs and 10,000 pounds for an OSHA hazardous chemical. Keep in mind that if you equal or exceed these threshold quantities **at any time** during the year, then you are subject to this reporting requirement. The threshold quantities should **not** be considered **the average amount** of a given chemical onsite during the year.

EPA publishes two types of inventory forms, **Tier I** and **Tier II**, for reporting this information. The Tier I form requires facilities to report general information on the amount and location of hazardous chemicals. Tier II forms require more detailed information on each hazardous chemical. At a minimum, you must report the information contained in EPA's Tier I form.

As required by statute, Tier I information includes the general elements listed below:

- An estimate (in ranges) of the maximum amount of chemicals for each hazard category (i.e., immediate health, delayed health, fire, sudden release, and reactive) present at the facility at any time during the preceding calendar year;

- An estimate (in ranges) of the average daily amount of chemicals in each category; and

•• The general location of hazardous chemicals in each category.

While federal regulations require only the submission of a Tier I form, EPA encourages, and some states require, the use of the Tier II form. EPA offers assistance in completing the Tier II form through its *Tier2 Reporting and Inventory System*. This system walks you through the preparation of the Tier II reporting form. For more information, access **http://www.epa.gov/swercepp/tools.html/**.

> *Some states have their own form and may allow electronic reporting. Contact your state for more information.*

7.5 Toxic Chemical Release Reporting– Section 313

Section 313 of EPCRA requires certain designated businesses to submit annual reports (commonly referred to as Form Rs and Form As) on the amounts of more than 600 EPCRA Section 313 chemicals and chemical categories released and otherwise managed (40 CFR 372). EPA selects the chemicals based on the potential for acute health effects, chronic health effects, and environmental effects. The original list of chemicals subject to Section 313 reporting was a combination of chemical lists from the states of New Jersey and Maryland.

All facilities meeting the Section 313 reporting criteria must report the annual releases and/or other waste management activities (routine and accidental) of EPCRA Section 313 chemicals to all environmental media. A separate report is required for each listed chemical that is manufactured (including imported), processed or otherwise used above the reporting threshold. The reports must be submitted to EPA and State or Tribal governments, on or before July 1, for activities in the previous calendar year. The owner/operator of the facility on July 1 is primarily responsible for the report, even if the owner/operator did not own the facility during the reporting year.

EPA can modify the list of chemicals, or industry or the public can petition EPA to modify the list. Therefore, before completing your annual report, be sure to check the most **current** list included with the *Toxic Chemical Release Inventory Reporting Forms and Instructions (TRI Forms and Instructions)*. You can request this package from the Resource Conservation and Recovery Act/Underground Storage Tank (RCRA/UST), Superfund and Emergency Planning and Community Right-to-Know Act (EPCRA) Hotline at 1-800-424-9346 or 703-412-9810 (Washington, DC, metropolitan area).

7.5.1 EPCRA Section 313 Reporting Guidance for Food Processors

To assist food processing facilities in complying with the reporting requirements of EPCRA Section 313 and Section 6607 of the Pollution Prevention Act of 1990 (PPA), EPA's Office of Pollution Prevention and Toxics (OPPT) has prepared a guidance manual, entitled *EPCRA Section 313 Reporting Guidance for Food Processors* (EPA 745-R-98-011, September 1998). This new guidance supplements the *TRI Forms and Instructions*, and supercedes EPA's earlier document, entitled *Section 313 Emergency Planning and Community Right-to-Know Act, Guidance for Food Processors* (June 1990). Additional discussion on specific issues can be found in EPA's current version of *EPCRA Section 313, Questions and Answers*, which is available on EPA's TRI website (**http://www.epa.gov/opptintr/tri**), or by contacting the RCRA/UST, Superfund and EPCRA Hotline at 1-800-424-9346 or 703-412-9810.

The *EPCRA Section 313 Reporting Guidance for Food Processors (9/98)* includes the following changes or additions: 1) more detailed examples and common industry-specific reporting errors;

2) EPCRA 313 regulations promulgated since 1990; 3) EPA's interpretive guidance on various issues specific to the food processing industry; and 4) input from the National Food Processors Association and the Food Industry Environmental Council. The objectives of the guidance are to reduce the level of effort expended by those facilities that prepare an EPCRA Section 313 report, and to increase the accuracy and completeness of the data reported on Form Rs or Form As by the food processing industry.

OPPT's *EPCRA Section 313 Reporting Guidance* is an essential, industry-specific compliance assistance tool. Acquiring it should be a high priority for environmental managers in the food processing industry. The following sections of this multimedia compliance guide briefly summarize, excerpt, or cross-reference text, tables and industry-specific examples found in OPPT's new guidance for food processors. Consult OPPT's guidance for the wealth of detailed industry-specific examples and the discussions of common reporting errors and compliance issues.

7.5.2   EPCRA Section 313 Reporting Requirements

To understand EPCRA 313 reporting requirements, you must first understand how EPCRA defines the terms, "facility" and "establishment." The term facility is defined as "all buildings, equipment, structures and other stationary items which are located on a single site or on contiguous or adjacent sites and which are owned or operated by the same person (or by any person which controls, is controlled by, or is under common control with such person)." A facility may contain more than one "establishment." An "establishment" is defined as "an economic unit, generally at a single physical location, where business is conducted, or services or industrial operations are performed" (40 CFR 372.3).

**Common Error: Multi-Establishment Facilities
and Agricultural Operations**

Some multi-establishment food processing facilities overlook the fact that they may have to submit Form R or Form A reports for chemicals used in agricultural operations. (See *EPCRA Section 313 Reporting Guidance (9/98)*, pages 2-7 and 2-8 for further explanation.)

The following section briefly describes EPCRA Section 313 reporting requirements to help you determine if these requirements apply to your facility, and if yes, what kind of a report(s) (e.g., Form R or Form A) you should prepare. Note the standard report is Form R. However, to reduce the reporting burden for small businesses, EPA established an alternative threshold reporting level that is discussed later in this section. If your facility does not exceed this level and meets certain other criteria, then you may file Form A -- a Certification Form -- rather than Form R.

How do you determine if your facility must prepare an EPCRA Section 313 report? The answers to the following four questions will help you decide:

1) Is the SIC Code for your facility included in the list covered by EPCRA Section 313 reporting?

2) Does your facility employ 10 or more full time employees or their equivalent?

3) Does your facility manufacture (which includes importation), process, or otherwise use EPCRA Section 313 chemicals?

4) Does your facility exceed any applicable thresholds of EPCRA Section 313 chemicals (either 25,000 pounds per year for manufacturing, or 25,000 pounds per year for processing, or 10,000 pounds per year for otherwise use)?

If you answer "**No**" to any of the first three questions, you are **not** required to prepare any Form R or Form A reports. If you answer "**Yes**" to **all** of the first three questions, you must then address question four. To address question four, you must do the following: a) complete a threshold calculation for each EPCRA Section 313 chemical at your facility; and then, b) for each EPCRA 313 chemical exceeding a threshold, you must submit a Form R **or** Form A.

To get a clearer picture of the decision making process, refer to Figure 7-1 *EPCRA Section 313 Reporting Decision Diagram*. (This diagram is identical to the one found in the *EPCRA Section 313 Reporting Guidance (9/98)*, page 2-3.)

Question 1: SIC Code Determination

Facilities with certain SIC codes are covered by EPCRA 313 reporting requirements. These include SIC Codes shown in the table below (40 CFR 372.22). For assistance in determining which SIC code(s) best suits your facility, based on the activities onsite, refer to *Standard Industrial Classification Manual*, 1987, published by the Office of Management and Budget.[1]

---

[1]See *EPCRA Section 313 Reporting Guidance* (9/98), pages 2-4 and 2-5 for a discussion of SIC codes and codes of the North American Industry Classification System (NAICS). The NAICS is replacing the SIC system. Dual systems will be used for a transition period which began in 1997. The NAICS uses six digits (vs. four for the SIC) which allows for a finer division of industries in a larger economy. Additional information on the NAICS is available from the U.S. Census Bureau on http://www.census.gov/epcd/www/naics.html.

Figure 7-1. EPCRA Section 313 Reporting Decision Diagram

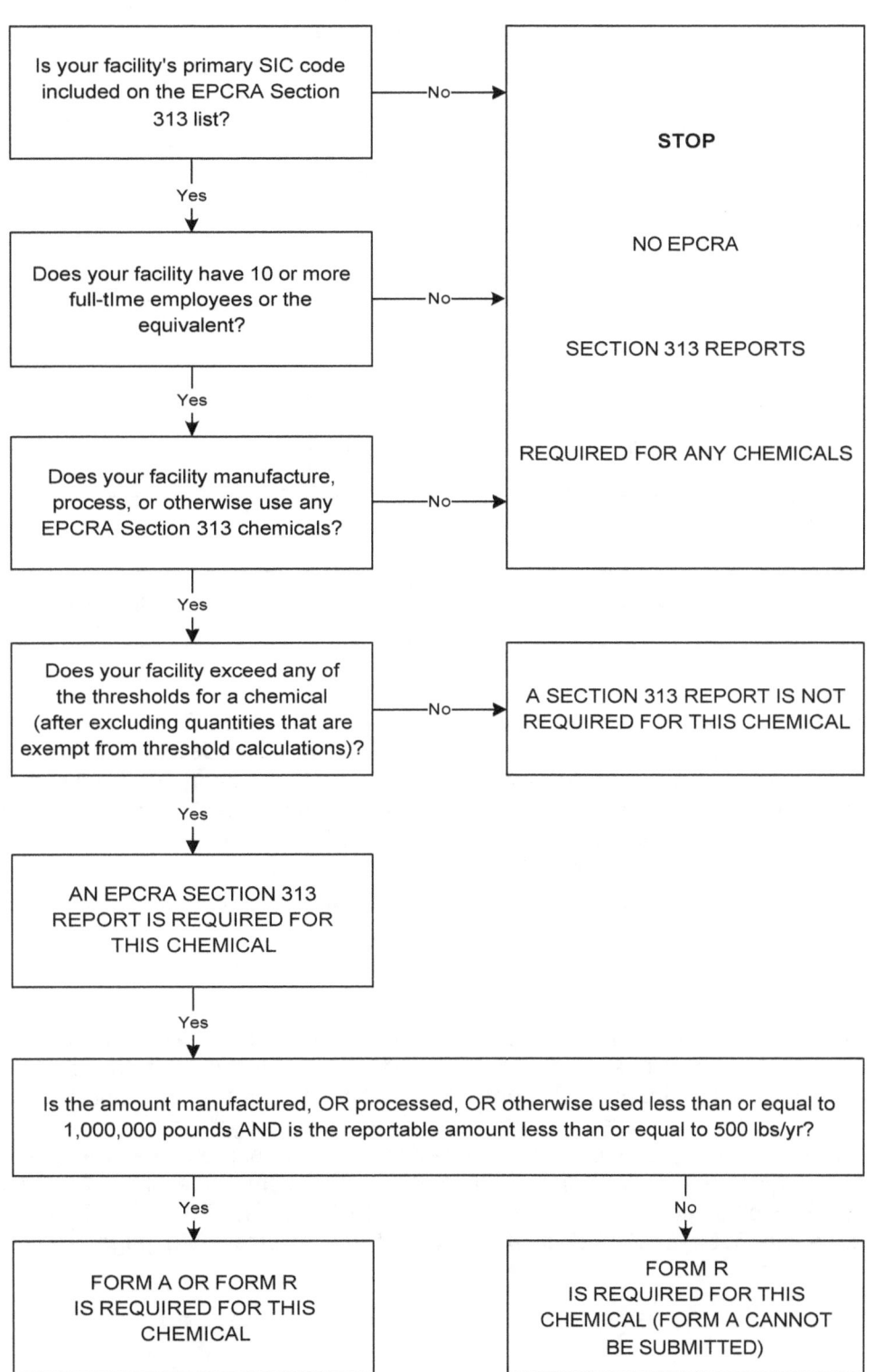

Table 7-2. SIC Codes Covered by EPCRA Section 313 Reporting

SIC CODE MAJOR GROUPS		
SIC Codes	**Industry**	**Qualifiers**
10	Metal Mining	Except SIC Codes 1011, 1081, 1094
12	Coal Mining	Except SIC Code 1241
20 through 39	Manufacturing	All SIC Codes
4911, 4931, and 4939	Electric and Other Services and Combination Utilities	Limited to facilities that combust coal and/or oil for the purpose of generating electricity for distribution in commerce
4953	Refuse Systems	Limited to facilities regulated under RCRA Subtitle C
5169	Chemicals and Allied Products	None
5171	Petroleum Bulk Stations and Terminals	None
7389	Business Services	Limited to facilities primarily engaged in solvent recovery services on a contract or fee basis.

Most food processing facilities are in SIC Major Group 20 (a covered SIC code). If a food processing facility meets the employee and chemical activity thresholds in addition to being in a covered SIC code, it is required to prepare a Form R (or Form A) Report. If your facility has more than one SIC code (i.e., several establishments with different SIC codes are owned or operated by the same entity and are located at your facility), then you must determine what is the **primary** SIC code for your facility according to criteria set up under EPCRA Section 313 requirements. (See *EPCRA Section 313 Reporting Guidance (9/98)*, pages 2-4 to 2-8.)

Question 2: Number of Employees

If your facility has 10 or more full-time employees or the equivalent, you are required to report provided that your facility also is in a covered SIC code and meets the chemical activity threshold for any EPCRA Section 313 chemical. A full time employee equivalent is defined as a work year of 2,000 hours. Therefore, if your facility's employees aggregate 20,000 or more hours in a calendar year, you meet the employee criterion of "10 or more employees or the equivalent." Remember to include any part time and seasonal employees in your calculations, including workers on an adjacent farm that are part of the facility (40 CFR 372.22). (Refer to the example presented in the *EPCRA Section 313 Reporting Guidance (9/98)*, page 2-10.)

Question 3: Chemical Activity Categories

If you answered "Yes" to Questions 1 and 2 above, then you must determine which EPCRA Section 313 chemicals are "manufactured," "processed," or "otherwise used" at your facility. You should prepare a list of all chemicals used by **all** establishments at the facility, including the chemicals found in mixtures and trade name products. You should compare your list to the **current** list of EPCRA Section 313 chemicals found in the *TRI Forms and Instructions* for the reporting year.

OPPT has prepared the following table of EPCRA Section 313 chemicals commonly reported for the food processing industry. The table has two columns. The first column lists the industrial process (water treatment, refrigerant uses, reactants, catalysts, etc.), and the second column lists examples of EPCRA Section 313 chemicals reported by this industry. This list is not all inclusive; therefore, you should use it only as a guide. (This table is identical to Table 2-3 in the *EPCRA Section 313 Reporting Guidance (9/98)*, page 2-11.)

Table 7-3. EPCRA Section 313 Chemicals Commonly Encountered in Food Processing

Process	Chemicals
Water Treatment	Chlorine and chlorine dioxide
Refrigerant Uses	Ammonia, ethylene glycol, Freon 113, dichlorodifluoromethane, CFC-114, chlorodifluoromethane
Food Ingredients	Phosphoric acid, various food dyes, various metals (e.g. zinc, copper, manganese, selenium, metal compounds) and peracetic acid
Reactants	Ammonia, benzoyl peroxide, chlorine, chlorine dioxide, ethylene oxide, phosphoric acid, propylene oxide
Catalysts	Nickel and nickel compounds
Extraction/Carrier Solvents	n-Butyl alcohol, dichloromethane, n-hexane, phosphoric acid, cyclohexane, and tert-butyl alcohol
Cleaning/Disinfectant Uses	Chlorine, chlorine dioxide, formaldehyde, nitric acid, phosphoric acid, and 1,1,1-trichloroethane
Wastewater Treatment	Ammonia, hydrochloric acid aerosols, and sulfuric acid aerosols
Fumigants	Bromomethane, ethylene oxide, propylene oxide, and bromine
Pesticides/Herbicides	Various pesticides and herbicides (e.g., aldrin, captan, 2, 4-D, hydrazine, lindane, maneb, parathion, zineb, malathion, atrazine, diazinon bromine, and naphthalene)

Table 7-3.   EPCRA Section 313 Chemicals Commonly
Encountered in Food Processing (continued)

Process	Chemicals
Byproducts	Ammonia, chloroform, methanol, hydrogen fluoride, and nitrate compounds
Can Making/Coating	Various ink and coating solvents (e.g. glycol ethers, MEK, toluene, methyl isobutyl ketone, xylene), various listed metals (e.g. manganese, nickel, chromium), and various metal pigment compounds (e.g., many pigments contain copper, barium, chromium, zinc, or lead)

Question 4: Threshold Determinations

After you identify the EPCRA Section 313 chemicals at your facility, then you must evaluate the activities involving each chemical, and determine if any of these activities meet any of the activity thresholds. EPCRA Section 313 reporting requirements define three activity categories for each EPCRA Section 313 chemical. These include "manufacturing" (which includes importing), "processing", and "otherwise using."

Brief definitions for the manufacturing (including importation), processing, and otherwise using appear in the table below. (This table is identical to Table 2-4 in the *EPCRA Section 313 Reporting Guidance (9/98)*, pages 2-12 and 2-13.)

The EPCRA Section 313 requirements divide each of these activity categories into subcategories. OPPT's guidance discusses each category and subcategory of activity along with relevant examples from the food processing industry. For more information, refer to the tables in Chapter 3 of the *EPCRA Section 313 Reporting Guidance (9/98)*, pages 3-8, 3-9 and 3-10. These tables are,

> Table 3-2 Definitions and Examples of Manufactured Chemicals
> Table 3-3 Definitions and Examples of Processed Chemicals
> Table 3-4 Definitions and Examples of Otherwise Used Chemicals.

Table 7-4. Activity Categories

Activity Category	Definition	
Manufacture	To produce, prepare, import, or compound a toxic chemical. "Manufacture" also applies to a toxic chemical that is produced coincidentally during the manufacture, processing, otherwise use, or disposal of another chemical or mixture of chemicals as a byproduct, and a toxic chemical that remains in that other chemical or mixture of chemicals as an impurity during the manufacturing, processing, or otherwise use or disposal of any other chemical substance or mixture. An example would be the production of ammonia or nitrate compounds in a wastewater treatment system.	25,000
Process	To prepare a listed EPCRA Section 313 chemical, or a mixture or trade name product containing an EPCRA Section 313 chemical, for distribution in commerce (usually the intentional incorporation of an EPCRA Section 313 chemical into a product). For example, zinc compounds may be processed as an additive in dog food, and would have to be reported if you exceeded the reporting threshold. Processing includes the preparation for sale to your customers (and transferring between facilities within your company) of a chemical or formulation that you manufacture. For example, if you manufacture a chemical or product, package it, and then distribute it into commerce, this chemical has been manufactured AND processed by your facility.	25,000
Otherwise Use	Generally, use of a listed EPCRA Section 313 chemical that does not fall under the Manufacture or Process definitions is classified as Otherwise Use. A listed chemical that is Otherwise Used is not intentionally incorporated into a product that is distributed in commerce, but may be used instead as a manufacturing or processing aid (e.g., catalyst), in waste processing, or as a fuel (including waste fuel). For example, n-butyl alcohol used as a carrier solvent for spices is classified as Otherwise Used. On May 1, 1997 U.S. EPA revised the interpretation of "otherwise use". The following new "otherwise use" definition becomes effective with the 1998 reporting year (62 FR 23834, May 1, 1997). Otherwise use means "any use of a toxic chemical contained in a mixture or other trade name product or waste, that is not covered by the terms "manufacture" or "process." Otherwise use of a toxic chemical does not include disposal, stabilization (without subsequent distribution in commerce), or treatment for destruction unless: 1) The toxic chemical that was disposed, stabilized, or treated for destruction was received from off site for the purposes of further waste management; OR 2) The toxic chemical that was disposed, stabilized, or treated for destruction was manufactured as a result of waste management activities on materials received from off site for the purposes of further waste management activities."	10,000

Associated with each activity category is an activity threshold summarized in the next table. These thresholds have been in effect since the reporting year of 1989. The activity thresholds apply to each EPCRA Section 313 chemical. Note that the threshold determination for each of the three activity categories is mutually exclusive of the others. Therefore, you must conduct a separate threshold determination for each chemical for each activity category. If you exceed any one of the activity thresholds, then you must submit a Form R (or Form A) report.

Table 7-5. EPCRA Section 313 Reporting Activities/Thresholds

Chemical Activity	Activity Threshold
Manufacturing	25,000 pounds/year
Processing	25,000 pounds/year
Otherwise Use	10,000 pounds/year

The threshold determination is based **solely** on the quantity **actually** manufactured (including imported), processed, or otherwise used, **not** on the quantity of chemicals stored onsite or purchased. Therefore, EPCRA Section 313 chemicals that are bought and stored, but are not incorporated into a product for distribution **or** not otherwise used onsite during the reporting year, are not counted towards any activity thresholds.

Many EPCRA Section 313 chemicals are present as impurities or as small components of mixtures. These quantities must also be considered in threshold determinations unless the concentration is below the *de minimis* value. In some cases, if a chemical is present below *de minimis* concentration, it may be exempt. See OPPT's guidance (9/98), pages 3-10 to 3-18, for more information on how to evaluate *de minimis* and three other classes of exemptions, including article, facility-related, and activity-related exemptions.

Several chemicals on the EPCRA Section 313 chemical list include qualifiers related to use or form (e.g., fume or dust, solutions, acid aerosols, etc.). Some chemicals are reportable **only if** manufactured by a specific process or in a specified activity category. OPPT's *EPCRA Section 313 Reporting Guidance (9/98)*, pages 3-5 to 3-8, contains an industry-specific discussion of these qualifiers, the associated chemicals and how these typically apply to the food processing industry. A detailed discussion of the qualifier criteria can be found in the *TRI Forms and Instructions*.

To determine if a chemical exceeds a reporting threshold, you must calculate the annual activity usage of that chemical. For example, start with the amount of the chemical at the facility as of January 1; add any purchases during the year and the amount manufactured (included imported); and subtract the amount left in the inventory on December 31. If necessary, adjust the total to account for exempt activities. Then compare the result to the appropriate activity threshold to determine if you are required to submit an EPCRA Section 313 Form R report for that chemical. OPPT's guidance (pages 3-22 and 3-23) provides a blank worksheet and a sample illustration to assist you with threshold calculations.

7.5.3 How to Estimate Releases and Other Waste Management Amounts

Your must file a Form R report for **each** EPCRA Section 313 chemical if that chemical exceeds any activity threshold for manufacturing, **or** processing, **or** otherwise use (provided that you also meet the SIC code and employee criteria). However, you may be eligible to file a Form A certification statement, rather than a Form R, provided that you meet certain criteria described below.

The Form R consists of the following two parts:

Part I, Facility Identification Information. Except for the signature, this part may be photocopied and re-used for each Form R you submit. Each Form R must have an original signature.

Part II, Chemical Specific Information. You must complete this part separately for each EPCRA Section 313 toxic chemical or chemical category. Among other items of information in Part II, you must provide the total annual reportable amount. The **reportable amount** is defined as the sum of the onsite amounts released (including disposal), treated, combusted for energy recovery and recycled, combined with the sum of the amounts transferred offsite for recycling, energy recovery, treatment, and/or release (including disposal). This total corresponds to the total of data elements 8.1 through 8.7 on the 1997 version of the Form R. Note: You **cannot** re-use this portion year after year, even if reporting has not changed.

The Form A, also referred to as the "Certification Statement," is an alternative to Form R. Form A first became available in reporting year 1994. EPA developed Form A (59 FR 61488, November 1994) to reduce the annual reporting burden for facilities that meet both of the following criteria:

- Chemical Activity Thresholds: The amount of the EPCRA Section 313 chemical manufactured, or processed, or otherwise used must not exceed one million (1,000,000) pounds. [Note: The threshold determination for each activity category is mutually exclusive of the others; i.e., each threshold must be evaluated independently. Therefore, if the quantity for any one activity threshold **exceeds 1,000,000 pounds**, then your facility **cannot** submit Form A.]

  And

- Annual Reportable Amount: The total annual reportable amount of the EPCRA Section 313 chemical **cannot exceed five hundred (500) pounds** per year. As stated above, the **reportable amount** is defined as the sum of the on site amounts released (including disposal), treated, combusted for energy recovery and recycled, combined with the sum of the amounts transferred off site for recycling, energy recovery, treatment, and/or release (including disposal). This total corresponds to the total of data elements 8.1 through 8.7 on the 1997 version of the Form R.

The Form A Certification Statement must be submitted for each eligible EPCRA Section 313 chemical. Like the Form R, Form A includes facility identification information. However, Form A does not require your facility to report any estimate of releases and other waste management quantities. Rather, your facility must simply certify that the total annual reportable amount does not exceed 500 pounds for that particular chemical.

For industry-specific assistance in calculating reportable amounts, consult Chapter 4 "Estimating Releases and Other Waste Management Quantities" of OPPT's *EPCRA Section 313 Reporting Guidance (9/98)*. This chapter provides a detailed, step-by-step discussion of how to calculate the release and other waste management amounts for any Section 313 chemical for which your facility must submit a report. This procedure consists of:

- Preparation of a detailed process flow diagram;

- Identification of potential sources of toxic chemicals released and/or otherwise managed;

- Identification of the potential types of releases and/or other waste management activities from each source; and

- Determination of the most appropriate method(s) for estimating the quantities of listed toxic chemicals and/or other waste management activities.

Chapter 4 of OPPT's guidance also briefly analyzes twelve chemical use categories commonly found in the food processing industry. For each of these twelve categories, the guidance does the following: lists the commonly used EPCRA Section 313 chemicals; gives an overview of the process involved; identifies the appropriate chemical activity category(ies) and reporting thresholds; describes methods for estimating quantities of chemicals released and otherwise managed as waste; and discusses common reporting errors.

Consult *TRI Forms and Instructions* for detailed directions on how to prepare and submit a Form R or a Form A report for **each** listed EPCRA Section 313 chemical. You have the option of submitting Form R(s) electronically via EPA's Automated Toxic Chemical Release Inventory Reporting Software (ATRS). EPA encourages the use of ATRS to save you time in data entry and photocopying, and to reduce errors by means of the online validation routines and use of pick lists within the software.

The ATRS can be found on the Internet at **http://www.epa.gov/opptintr/atrs**. It is available in both DOS and Windows versions. Call the ATRS User Support Hotline at 703-816-4434 for more information.

7.5.4   EPCRA Section 313 Recordkeeping

Complete and accurate records are absolutely essential to meaningful compliance with EPCRA Section 313 reporting requirements. Compiling and maintaining good records will help you to reduce the effort and cost in preparing future reports, and to document how you arrived at the reported data in the event of an EPA compliance audit. EPA requires you to maintain records substantiating the Form R or Form A submission, for a minimum of three years. Each facility must keep copies of the Form R or Form A along with all supporting documents, calculations, work sheets, and other forms that you use to prepare the Form R or Form A. EPA may request this supporting documentation during a regulatory audit.

Specifically, EPA requires that the following records must be maintained for a period of three years from the date of the submission of a report (summarized from 40 CFR 372.10):

1)   A copy of each report that is submitted.

2)   All supporting materials and documentation used by the person to make the compliance determination that the facility or establishment is a covered facility.

3)   Documentation supporting the report that is submitted, including documentation supporting:

   •• Claimed allowable exemptions;
   •• Threshold determinations;
   •• Calculations for each quantity reported as being released, either on or off site, or otherwise managed as waste;
   •• Activity use determinations, including dates of manufacturing, processing, or use;
   •• The basis of all estimates;
   •• Receipts or manifests associated with transfers to off-site locations; and
   • Waste treatment methods, treatment efficiencies, ranges of influent concentrations to treatment, sequential nature of treatment steps, and operating data to support efficiency claims.

4)   All supporting materials used to make the compliance determination that the facility or establishment is eligible to submit a Form A.

5)   Documentation supporting the Form A, including:

   •• Data supporting the determination that the alternate threshold applies;
   •• Calculations of annual reporting amounts; and
   • Receipts or manifests associated with the transfer of each chemical in waste to offsite locations.

Because EPCRA Section 313 reporting does not require additional testing or monitoring, you must determine the best readily available source of information for all estimates. Some facilities may have detailed monitoring data and offsite transfer records that are used for estimates, while others may only use purchase and inventory records. Examples of records that you should keep, if applicable, might include:

   ••   Each Form R or Form A submitted;
   ••   EPCRA Section 313 Reporting Threshold Worksheets (sample worksheets can be found in Chapter 3 of this document as well as in the *TRI Forms and Instructions*);

- • Engineering calculations and other notes;
- • Purchase records from suppliers;
- • Inventory data;
- • National Pollutant Discharge Elimination System (NPDES)/State Pollutant Discharge Elimination System (SPDES) permits and monitoring reports;
- • EPCRA Section 312 Tier II reports;
- • Monitoring records;
- • Air permits;
- • Flow measurement data;
- • RCRA hazardous waste generator's reports;
- • Pretreatment reports filed with local governments;
- • Invoices from waste management firms;
- • Manufacturer's estimates of treatment efficiencies;
- • Comprehensive Environmental Response, Compensation and Liability Act (CERCLA) reportable quantity (RQ) reports;
- • RCRA manifests; and
- Process flow diagrams (including emissions, releases, and other waste management activities).

SECTION 8 CONTENTS

8. HOW DO I COMPLY WITH THE HAZARDOUS WASTE REGULATIONS?

8.1   Introduction

As a food processor, you produce wastes that could be hazardous. Therefore, it is important that you identify and manage them properly to protect yourself, coworkers, and others in your community, as well as the environment. As the waste generator, you are responsible for all steps in hazardous waste management, from generation to storage to final disposal. **You can be held liable for any mismanagement of your wastes, even after they leave your facility. So, it is important for you to know the requirements.**

This section explains the hazardous waste law, known as the Resource Conservation and Recovery Act (RCRA), and its regulations which impose requirements on how you , store, must handle and dispose of the wastes you generate in your food processing facility. Sections 8.5 *Compliance Requirements for CESQGs and* Section 8.6 *Compliance Requirements for SQGs and LQGs* focus on the major federal requirements with which you must comply.

In some instances, the states impose additional and more stringent requirements on how you handle your wastes. It is critical, therefore, that you review you state's requirements and contact your state hazardous waste agency for any additional requirements.

If your facility has an underground storage tank (UST) system, you are subject to RCRA Subtitle I requirements. Section 8.7 *Underground Storage Tanks* provides an overview of these requirements, which pertain to USTs containing petroleum products or substances defined as hazardous under The Comprehensive Environmental Response, Compensation, and Liability Act (CERCLA). If your facility has aboveground storage tanks (ASTs), see Section 4.6 *How Do I Comply with Oil Pollution Prevention Regulations* for more information?

Here are the steps you should follow to ensure compliance with this law:

- Determine whether you have hazardous waste at your food processing facility
- Determine your hazardous waste generator status
- Meet the major requirements based on your hazardous waste generator status.

Your ability to comply with RCRA regulations depends on your understanding of what constitutes a hazardous waste. This definition is fundamental because it determines how wastes must be managed. It is important to recognize that the definition of hazardous waste is not straightforward.

8.2 What is Hazardous Waste?

The answer to this question is complex and requires you to follow several steps. First you must determine what types of wastes your facility generates and how each waste will be managed. RCRA defines two categories of wastes, solid and hazardous. Hazardous waste is a subset of solid waste. Therefore, if your waste does not meet the definition of solid waste, it will not be hazardous waste by definition.

8.2.1 Solid Waste

The definition of solid waste is so broad that most materials you dispose of fall within it. Under the RCRA statute, a "solid waste" is defined as any solid, liquid, or contained gaseous material that is discarded by being disposed of, burned or incinerated, or recycled (and not excluded under the regulations). The first step to determine whether or not you are generating a solid waste is to identify which of the categories of "secondary materials" your waste fits. These are:

> *See Section 10.2 for information on managing nonhazardous, solid waste under RCRA Subtitle D.*

- Spent material
- Scrap metal
- Listed or characteristic by-product
- Listed or characteristic sludge
- Commercial chemical product.

Specific definitions of each of these can be found in the *Code of Federal Regulations* (CFR) at 40 CFR 261.1. For additional guidance in classifying your wastes according to these categories, see the January 4, 1985 *Federal Register*.

The second step is to identify how you plan to manage the waste. The classification as a "solid waste" will depend on whether you plan to dispose of the waste, burn it for energy recovery, incinerate it, or recycle it, as well as how long you plan to store it prior to recycling it. The most complex questions regarding the definition of solid waste arises in the context of recycling activities. If you plan to recycle your waste, and you have any uncertainty concerning its classification, you should consult your state regulatory agency, EPA regional office, or the RCRA/UST, Superfund and EPCRA hotline at 1-800-424-9346 or 703-412-9810.

There are a number of disposable materials that are excluded from the definition (in RCRA) of solid waste (and thus hazardous waste) including:

- Domestic sewage and any mixture of domestic sewage and other wastes that pass through a sewer system to a publicly owned treatment works (POTW) for treatment. "Domestic sewage" means any untreated sanitary wastes that pass through a sewer system.

- Industrial wastewater discharges that are point source discharges (i.e., they are discharged from a single point or pipeline) and are regulated under Section 402 of the Clean Water Act (CWA). This exclusion only applies to the actual point source discharge. Industrial wastewaters that are being collected, stored, or treated before discharge are not excluded, nor are sludges that are generated by industrial wastewater treatment. If you are treating your own wastewater, the sludges are considered solid waste, and could be hazardous waste.

8.2.2 Hazardous Waste

Once you have determined that your material to be disposed of is solid waste, you then must determine if it is hazardous. For a waste to be classified as hazardous, it either:

- Is on one of the four lists of hazardous wastes (see *Listed Wastes* below) published in the federal RCRA regulations (40 CFR 261),

- Demonstrates one or more of the four hazardous waste characteristics of ignitability, corrosivity, reactivity, or toxicity (see *Characteristic Wastes* below), **or**

- Is a **mixture** of a listed hazardous waste and other wastes. It is important to note that waste mixtures that include hazardous wastes are regulated as hazardous waste regardless of the proportions of the mixture.

Listed Hazardous Wastes

Your waste is considered a hazardous waste if it appears on one of four lists (see table below) published in the hazardous waste regulations (40 CFR 261 Subpart D). Currently, more than 400 wastes are on these lists. Wastes are listed as hazardous because they are known to be harmful to human health and the environment when not properly managed. Even when properly managed, some listed wastes are so dangerous that they are called **acutely hazardous wastes**. Examples of acutely hazardous wastes include wastes generated from some pesticides that can be fatal to humans or animals in low doses.

Each list represents a different category of hazardous wastes and has a different alphabetical letter (F, K, U, and P). The categories are defined by the source of the waste.

List:	Listed hazardous wastes include:	Wastes generated in food processing:
F	The F list (40 CFR 261.31) designates as hazardous particular wastes from certain common industrial or manu-facturing processes. Because the processes producing these wastes can occur in different sectors of industry, the F list wastes are known as wastes from nonspecific sources (e.g., degreasing)	Food processors will most likely generate spent solvent wastes, which are F-listed wastes F001-F005.
K	The K list (40 CFR 261.32) designates as hazardous particular wastestreams from certain specific sectors of industry. K list wastes are known as wastes from specific sources.	Food processors typically do not generate these types of wastes.

List:	Listed hazardous wastes include:	Wastes generated in food processing:
U and P	The U and P lists are similar in that both list as hazardous pure or commer-cial grade formulations of certain specific unused chemicals. P wastes are all acutely hazardous (40 CFR 261.33).	Food processors may generate these types of wastes (e.g., unused pesticide of pure heptachlor [P059]).

If a waste is not found in any of these four federal lists, it still may be on a state hazardous waste list. For example, many states list waste petroleum oil as a hazardous waste.

Characteristic Wastes

If a waste does not appear on one of the EPA lists discussed above, it may still be considered a hazardous waste if it has one or more of the following characteristics:

- It can readily catch fire and sustain combustion. This is called an **ignitable** waste (40 CFR 261.21). EPA selected a flashpoint test as the method for determining whether a liquid waste is combustible enough to deserve regulation as hazardous. A non-liquid waste is only ignitable

    > *The flashpoint test determines the lowest temperature at which a chemical ignites when exposed to flame.*

    if it can spontaneously catch fire under normal handling conditions and can burn so vigorously that it creates a hazard. Ignitable wastes carry the waste code D001. Examples are paint wastes, certain degreasers, or other solvents.

- It is an acidic or alkaline (basic) waste which can readily corrode or dissolve flesh, metal, or other materials. This is called a **corrosive** waste (40 CFR 261.22). EPA uses two criteria to identify corrosive hazardous wastes. The first is a pH test. Wastes with a pH • 12.5 or • 2 are corrosive under RCRA rules. A waste may also be corrosive if it has the ability to corrode steel in a specific EPA-approved test protocol. Corrosive wastes carry the waste code D002. Examples are waste rust removers, waste acid or alkaline cleaning fluids, and waste battery acid.

- It readily explodes or undergoes violent reactions. This is known as a **reactive** waste (40 CFR 261.23). Reactive hazardous wastes are relatively uncommon, and, in many cases, there is no reliable test method to evaluate a waste's potential to explode or react violently. Therefore, EPA uses a narrative criteria to define most reactive wastes.

    Under RCRA, a waste is reactive if it meets any of the following criteria:
    – It can explode or violently react when exposed to water or under normal handling conditions;
    – It can create toxic fumes or gases when exposed to water or under other conditions (e.g., heat or pressure); or
    – It meets the criteria for classification as an explosive under U.S. Department of Transportation (DOT) rules.

Wastes exhibiting the characteristic of reactivity are assigned the waste code D003. Examples are waste bleaches and other waste oxidizers.

- It is harmful or fatal when ingested or absorbed, or it leaches toxic chemicals into the soil or groundwater when disposed of on land. This is called a **toxic** waste (40 CFR 261.24). You can determine if your waste is toxic by having it tested using the Toxicity Characteristic Leaching Procedure (TCLP). If the waste contains any of the regulated contaminants at concentrations equal to or greater than the regulatory levels, then the waste exhibits the toxicity characteristic. The toxic waste carries the waste code associated with the constituent which exceeded the regulatory level.

> *The Toxicity Characteristic Leaching Procedure (TCLP) replicates the leaching process and other effects that occur when wastes are buried in a typical municipal landfill.*

Mixture Rule and Derived From Rule

The mixture and derived from rules operate differently for listed wastes and characteristic wastes. The mixture rule for listed wastes states that a mixture made up of any amount of a nonhazardous solid waste and any amount of a listed hazardous waste is considered a hazardous waste. In contrast, a mixture involving a characteristic waste is hazardous only if the mixture itself exhibits a characteristic.

The derived from rule governs the regulatory status of materials that are created by treating or changing a hazardous waste in some way. For example, ash created by burning a hazardous waste is considered **derived from** that hazardous waste. The derived from rule for listed wastes states that any material derived from a listed hazardous waste is also a listed hazardous waste. A treatment residue and materials derived from characteristic hazardous wastes are hazardous only if they themselves exhibit a characteristic.

Hazardous Waste Codes

Specific hazardous waste types have designated waste codes. A waste code is a four-digit classification system used by EPA to identify hazardous wastes on labels, shipping papers, and other records. All federal hazardous waste codes begin with a letter and are followed by numbers. All federal **listed** wastes begin with either "F", "K", "U", or "P"; **characteristic** wastes begin with the letter "D." For a complete listing of hazardous waste codes, consult 40 CFR 261. Many states have listed waste numbers that begin with the two-letter state abbreviation followed by two specific numbers that identify the state-listed waste. In order to determine what the waste code is for your hazardous waste, you need to look at the regulations. You should call your state environmental agency to determine where you can obtain a copy of your state's regulations.

8.2.3 Universal Waste

EPA issued the Universal Waste Rule in 1995 as an amendment to RCRA. It provides an alternative and less stringent set of management standards to those in the hazardous waste regulations (40 CFR 260-272) for three specific, but widely generated, types of waste that potentially would be regulated as hazardous. These wastes are:

- Batteries that are spent and will not be reclaimed or regenerated either at your facility or at a battery recycling/reclamation facility (under 40 CFR 266 Subpart G). Types of batteries that your facility may generate that would be universal wastes include those in electronic equipment, mobile telephones, portable computers, and emergency backup lighting.

- Pesticides that have been suspended or canceled including those that are part of a voluntary or mandatory recall under the Federal Insecticide, Fungicide, and Rodenticide Act (FIFRA) or by the pesticide registrant; are unused but managed as part of a waste pesticide collection program; or are obsolete or damaged.

- Mercury thermostats including temperature control devices containing metallic mercury.

The Universal Waste rule establishes requirements applicable to four types of universal waste generators or collectors: small quantity handlers, large quantity handlers, transporters, and destination facilities. Specific requirements of the universal waste rule can be found at 40 CFR 273.

8.3 Are My Wastes Hazardous?

Do you generate hazardous wastes at your facility? Since you are a food processor, the answer will probably be "yes." For your facility, the hazardous waste identification process involves the following steps, in this order:

> *Under RCRA, it is your responsibility to determine if a waste produced or generated at your facility is hazardous (40 CFR 262.11).*

- Complete an inventory of all wastes that are generated by your facility (see Section 3.0).

- For each waste, determine whether the material in question is a "solid waste" (see Section 8.2.1).

- Determine whether the solid waste in question is excluded from regulation.

- Determine whether the solid waste in question is a hazardous waste. You may need to send a sample of your waste to a laboratory for them to determine if the waste is hazardous (see Section 8.2.2).

- Make a list of all the hazardous waste you have at your facility and determine the waste code for each (see 40 CFR 261).

- If you are unsure, get help. For assistance, call EPA or your state environmental agency, a consultant, a licensed transporter, or the RCRA/UST, Superfund and EPCRA hotline at 703-412-9810 or 1-800-424-9346.

As discussed in Section 3.0, hazardous wastes are generated during many food processing operations. Table 3-2 *Waste Analysis for SIC Code 203 Facility* (see Section 3.0) presents some of the typical hazardous wastes generated by food processing facilities. While this list is not all inclusive, it gives you a general idea of the kinds of wastes generated for each operation that are considered to be hazardous.

With your inventory list, you may be able to determine which wastes are hazardous based on knowledge alone. For example, you may use a cleaning solvent that you know is a listed hazardous waste. For wastes whose regulatory status cannot be determined by knowledge alone, the appropriate analysis should be performed in order to determine if any suspected characteristics meet the federal criteria for definition of hazardous wastes.

An important *first* source of information about the chemicals you use is the material safety data sheet (MSDS). The MSDS will identify specific chemical properties of a material, such as whether a material is highly acidic or basic, whether solvents are present, and other chemical properties, such as ignitability (flashpoint). MSDSs will not provide you with all of the answers to your environmental questions, but they can help you identify specific characteristics of your hazardous waste. The MSDS is provided by your chemical supplier and gives general health and safety information about handling these chemicals. To ensure that your MSDSs are current, require your vendors to automatically supply you with an MSDS for new products and to have anyone approving purchases in your business ask for them from the supplier.

If MSDSs do not provide you with enough information to make your determination, consult the RCRA regulations in 40 CFR 261 or call the RCRA/UST, Superfund and EPCRA hotline for assistance at 703-412-9810 or 1-800-424-9346.

8.4 What is My Hazardous Waste Generator Category?

If you are a food processor and your operations cause hazardous waste to be generated, you must now determine your generator category (40 CFR 261). Your generator category is determined by the amount of hazardous waste that you generate each month. There are three federal categories of hazardous waste generators:

- Conditionally exempt small quantity generator (CESQG)
  (• 220 pounds (100 kg) of hazardous waste; or • 2.2 pounds (1 kg) acutely hazardous waste per month).

- Small quantity generator (SQG)
  (> 220 pounds (100 kg) and < 2,200 pounds (1,000 kg)of hazardous waste per month.

- Large quantity generator (LQG)

(• 2,200 pounds (1,000 kg) of hazardous waste; or > 2.2 pounds (1 kg) acutely hazardous waste; or > 220 pounds spill residue from acutely hazardous waste).

Hazardous waste generators are divided into these three categories, depending on the quantity of hazardous waste generated each month and on the cumulative amount of hazardous waste stored at your food processing facility at any time. The measured amount (by weight) of hazardous waste generated at your facility per calendar month determines which hazardous wastes requirements and standards apply to you.

Determining Your Generator Category

To determine which category applies to your facility, you must count all quantities of listed and characteristic hazardous wastes. These include wastes that are: (1) generated and collected at your facility prior to treatment or disposal; and (2) packaged and transported offsite.

Many hazardous wastes are liquids and are measured in gallons, not pounds. To approximate the number of pounds of liquid you have, multiply the number of gallons by 8.3 (because a gallon of water weighs 8.3 pounds, and many liquids have a density similar to water). Most MSDSs list the density or specific gravity of the product.

> **Rough Guide**
> * 27 gallons (about half of a 55-gallon drum) of waste with a density similar to water weighs about 220 pounds (100 kg).
> * 270 gallons of waste with a density similar to water weighs about 2,200 lbs (1,000 kg).

Keep in mind that you do <u>NOT</u> have to count the following:

* Wastes that are left on the bottom of containers that have been emptied by conventional means (i.e., pouring or pumping) and where no more than 2.5 cm (1 inch) of residue remains in the bottom of the container <u>or</u> no more than 3 percent by weight of the total capacity of the container remains in the container if the container is less than or equal to 110 gallons in size.

* Residues in the bottom of storage tanks, if the residue is not removed (i.e., residues left in the bottom of the storage container are not counted as long as they are not removed when the tank is refilled).

* Wastes that are reclaimed continuously onsite without storing the waste prior to reclamation.

* Wastes that you have already counted once during the calendar month, and treated onsite or reclaimed in some manner and used again.

* Wastes that are directly discharged to a municipal treatment plant or POTW without being stored or accumulated first. **This discharge to a POTW must comply with the CWA and any local POTW regulations (see Section 4.4 *Am I An Indirect Discharger?*).**

- Batteries, pesticides, and mercury thermostats which are regulated under the Universal Waste Rule (40 CFR 273), or lead-acid batteries to be reclaimed (40 CFR 266).

- Waste oil that meets the criteria for used oil and is to be managed and handled as used oil (40 CFR 279).

- Scrap metal that is recycled (40 CFR 261.6(a)(3)).

Table 8-1 illustrates the federal generation rates and storage time limits applicable to the generator categories. It is important to note that states may specify different categories than those specified in the federal regulations and you must meet the requirements specified in your state. For example, a state may only have two categories of generators; small and large, and have specific requirements for each of these generator categories.

From Table 8-1, you can see it pays to be in one of the SQG categories. There is more leeway for storage time, which will allow you to more cost-effectively manage your smaller quantities of hazardous waste. In addition, pollution prevention (P2) can help you change your generator status (see Section 11.0 *Pollution Prevention Techniques*).

Table 8-1. Federal Categories of Hazardous Waste Generators and Storage Time Limits Allowed

Generator Category	Monthly Hazardous Waste Generation Rate	Storage Time Limits
Conditionally Exempt Small Quantity Generator (CESQG)	• 220 pounds (100 kg); or • 2.2 pounds (1 kg) acute	No Limit
Small Quantity Generator (SQG)	> 220 pounds (100 kg) and < 2,200 pounds (1,000 kg); or • 2.2 lb (1 kg) acute	• 180 days or • 270 days if waste treatment/disposal facility is >200 miles away
Large Quantity Generator (LQG)	• 2,200 pounds (1,000 kg); or > 2.2 pounds (1 kg) acute; or • 220 pounds spill residue from acute	• 90 days

8.5 Compliance Requirements for CESQGs

Your food processing facility is considered a CESGQ if it consistently generates less than 220 lbs (100 kg) of hazardous wastes per month, and less than 2.2 lbs. (1 kg) of acutely hazardous wastes per month. As a CESQG, your compliance requirements are quite simple. There are three basic hazardous waste management requirements that apply to CESQGs:

- Identify your hazardous and acutely hazardous wastes and know which wastes you generate that are hazardous. As explained in Section 8.3, there are several steps you should follow in your hazardous waste identification process. Please refer back to Section 8.3 for more information. If you are not sure of what you should do, get help. This may mean calling EPA or your state environmental agency, a consultant, a licensed transporter, or the RCRA/UST, Superfund and EPCRA hotline at 703-412-9810 or 1-800-424-9346.

- Do not generate more than 220 lbs (or 100 kg) per month of hazardous waste or more than 2.2 lbs (1 kg) per month of acutely hazardous waste (this includes any wastes you shipped offsite for disposal during that month); and never store more than 2,200 lbs (1,000 kg) of hazardous waste or 2.2 lbs of acutely hazardous waste for any period of time.

- Ensure proper treatment and disposal of your waste. For CESQGs, proper treatment and disposal of hazardous wastes are fairly simple. It involves ensuring the waste is shipped to one of the following facilities, or if you treat (e.g., solvent distillation) or disposed of your hazardous waste at your facility, ensure that your disposal facility is:

  – A state or federally regulated hazardous waste management treatment, storage, or disposal facility (if your waste is hazardous).

  – A facility permitted, licensed, or registered by a state to manage municipal or industrial solid waste.

  – A facility that uses, reuses or legitimately recycles the waste (or treats the waste prior to use, reuse, or recycling).

  – A universal waste handler or destination facility subject to the universal waste requirements (if you choose to follow the universal waste requirements, which you are not required to do as a CESQG, see below).

  CESQG Self-Transporting of Hazardous Wastes. CESQGs are allowed to transport their own wastes to the treatment or storage facility, unlike SQGs and LQGs who are required to use a licensed, certified transporter. While there are no specific RCRA requirements for CESQGs who transport their own wastes, DOT requires all transporters of hazardous waste to comply with all applicable DOT regulations. Specifically, DOT regulations require all transporters, including CESQGs, transporting hazardous waste that qualifies as DOT hazardous material to comply with EPA hazardous waste transporter requirements found in 40 CFR 263.

- CESQGs are not required by federal hazardous waste laws to train their employees on waste handling or emergency preparedness, but it is strongly advised. **Keep in mind that employees who are responding to releases of hazardous substances and hazardous waste are required to be trained under Occupational Safety and Health Administration's (OSHA's) Hazardous Waste Operations and Emergency Response (HAZWOPER) requirements (see 29 CFR 1910.120).**

You must comply with these requirements to retain your CESQG status, and remain exempt from the more stringent hazardous waste regulations that apply to SQGs and LQGs. However, it is recommended that you follow the waste storage and handling requirements for SQGs to minimize the possibility of any leaks, spills, or other releases that potentially could cause economic hardship to your facility. Table 8-2 provides a summary of the federal CESQG requirements. Please remember the requirements in Table 8-2 are the **minimum federal** requirements. States may have more stringent and/or different requirements. Contact your state hazardous waste agency for these requirements.

8.6 Compliance Requirements for SQGs and LQGs

If you determine, based on the amount of waste you generate, that you are an SQG or LQG, you must comply with the following requirements:

- Waste identification
- EPA identification number
- Accumulation and storage limits
- Container management
- Personnel training
- Hazardous waste shipment labeling and placarding
- Reporting and recordkeeping requirements
- Contingency planning, emergency procedures, and accident prevention.

Requirements for hazardous waste generators cover the storage and handling, treatment, and disposal of the waste, from generation to final disposal. Table 8-2 provides a summary of the federal SGQ and LQG requirements. Please remember the requirements in Table 8-2 are the **minimum federal** requirements. State governments may have more stringent and/or different requirements. Contact your state environmental agency for state requirements.

Waste Identification

As a generator, you must determine whether your waste is hazardous. As explained in Section 8.3, there are several steps you should follow in your hazardous waste identification process (40 CFR 261). Please refer back to Section 8.3 for more information. If you are not sure of what you should do, get help. This may mean calling EPA or your state environmental agency, a consultant, a licensed transporter, or the RCRA/UST, Superfund and EPCRA hotline at 703-412-9810 or 1-800-424-9346.

EPA Identification Number

Each SQG and LQG is required to obtain an EPA identification number. These 12-character identification numbers are part of a national database on hazardous waste activities. Some states also require conditionally exempt small quantity generators to have identification numbers. Furthermore, companies that transport hazardous waste and facilities that store, treat, or dispose of regulated quantities of hazardous waste generated by food processing facilities must also have EPA identification numbers.

Table 8-2. Summary of Federal Hazardous Waste Generator Requirements

	CESQG	SQG	LQG
EPA/state ID Number	Not federally required	Federally required	Federally required
Monthly Generation Limits (Weight)	• 220 lb (100 kg) or • 2.2 lb (1 kg) acute	>220 lb (100 kg) and <2,200 lb (1,000 kg) or • 2.2 lb (1 kg) acute	• 2,200 lb (1,000 kg) or >2.2 lb (1 kg) acute or • 220 lb spill residue from acute
Maximum Onsite Accumulation Limits (Weight)	• 2,200 lb (1,000 kg) or • 2.2 lb (1 kg) acute or • 220 lb spill residue from acute	• 13,200 lb (6,000 kg)	No limit
Maximum Onsite Time Limits for Storage	No limit	• 180 days or • 270 days if waste treatment/disposal facility is >200 miles	• 90 days
Container Management	Not federally required	Federally required	Federally required
Reporting and Recordkeeping			
Uniform Hazardous Waste Manifest	Not federally required	Federally required	Federally required
Exception Reports	Not federally required	Federally required	Federally required
Biennial Reports	Not federally required	Not federally required	Federally required
Land Disposal Restriction Notifications	Not federally required	Federally required	Federally required
Contingency Planning and Notification (including personnel training)	Not federally required	Basic procedures required; employees must know proper waste handling and emergency procedures	Written plan required; hazardous waste handling training program required for employees
Used/Waste Oil Management Standards	Federally required	Federally required	Federally required

ow to obtain a hazardous waste generator number:

- Call or write your state hazardous waste management agency or EPA regional office and ask for a copy of EPA Form 8700-12 "Notification of Regulated Waste Activity." You will be sent a booklet containing the two-page form and instructions for filling it out. Note that a few states use a form that is different from the EPA form. If you contact your state first, you will be sent the appropriate form to complete.

- Complete one copy of the form for each of your food processing facilities where you generate or handle hazardous waste. There is no fee associated with filling out this form. Each site or location will receive its own unique EPA identification number. **You must use this identification number on all hazardous waste shipping forms.**

- Make sure you fill the form out completely and correctly and sign the certification. Send the form to the address listed in the booklet you received with the form.

An EPA identification number is a unique number that applies to a particular food processing facility site or location. **If you move your food processing facility to another location, you must notify EPA or the state of the new location, submit a new form, and obtain a new EPA identification number.** If hazardous waste was previously handled at the new location, and it already has an EPA identification number, you will be assigned that number for your relocated food processing facility.

Onsite Accumulation and Storage Limits

Onsite accumulation (storage) limits are based on the total **weight** of hazardous waste that can be accumulated at any time at your food processing facility before it must be shipped offsite (40 CFR 262.34). Exceedance of the accumulation limits can cause a change in your generator status and, therefore, a change in the applicable regulatory requirements. Onsite accumulation weight limits and storage time limits for each generator status are presented in Table 8-2. Storage time is allowed so that you can accumulate enough hazardous waste onsite in order to make shipping it offsite for treatment or disposal economical.

Container Management

Your food processing facility can store hazardous waste in 55-gallon drums, tanks, or other suitable containers, but it must comply with rules intended to protect human health and the environment and reduce the likelihood of damages or injuries caused by leaks or spills (40 CFR 265). The following list summarizes the most significant requirements for managing containers of hazardous waste, regardless of their size:

- Establish and clearly mark an accumulation (storage) area for your hazardous waste. This is your designated onsite hazardous waste storage area and is a collection area for your entire facility. The length of time you can store hazardous waste in this area depends on your generator status. The type of area and marking requirements are set by your state. For storage tanks constructed after September

30, 1986, this area must have a containment system sufficient to hold spills and leaks. The amount of hazardous waste which can be stored onsite is shown in Table 8-2.

- Properly label and mark all containers of hazardous waste in your hazardous waste storage area. Clearly mark each container with the words "HAZARDOUS WASTE," and with the date the waste was first collected in that container. (Labels for this purpose may be available from the waste hauler or a trade association.) When your waste is shipped offsite, it is important that your transporter is aware of and complies with DOT placarding requirements for the truck used to haul your waste. Further, many states require additional labeling, such as a description of the contents of the container.

- You can accumulate up to 55 gallons of hazardous waste in properly labeled containers or drums at or near the various parts of your facility where the waste is generated. This is called *satellite accumulation*. Once 55 gallons have accumulated, satellite waste must be moved to your designated onsite hazardous waste storage area prior to shipment offsite.

- Containers in satellite accumulation areas must be clearly marked with the words "HAZARDOUS WASTE" or with other wording that identifies the contents of the container. Once the amount of waste in the container or drum reaches 55 gallons, it must be marked with the date it reaches that amount, and it must be moved to the designated onsite hazardous waste storage area within 72 hours (3 days). The operator of the process is responsible for this container or drum as long as it is kept separate from the designated storage area.

- Mark the EPA waste code on the drum. Although marking the EPA waste code on the drum is not required by federal regulations, it is required by most states and is highly recommended.

- Keep containers in good condition, handle them carefully, and replace any leaking ones. If a container is in poor condition, the waste must be transferred to a container in good condition.

- Use containers made of or lined with materials that will not react with the waste. Do not store hazardous waste in a container if it may cause ruptures, leaks, corrosion, or other failure.

- Do not throw away containers with product in them. If you have a container that has less than one inch of product or less than three percent of the total amount of product remaining, the container can be crushed, recycled, or thrown away. Otherwise, you must scrape out the product on the inside and properly manage it as hazardous waste. There is a federal requirement to triple rinse containers that have held acutely hazardous waste prior to considering the containers to be empty. Your state may also mandate this; please contact your state regulatory agency for more information.

- Keep containers closed except when adding or removing wastes. Remember, if a funnel remains in a drum, the drum is considered open. Do not handle or store a container in such a way that may rupture it or cause it to leak.

- Inspect the containers for leaks or corrosion every week. During your inspection, it is recommended that you make sure that:

  - All drums are labeled/marked appropriately

  - There is sufficient space to walk in the storage area and there is required space (36 inches) between rows of drums

  - All drums are stacked properly

  - All drum lids are closed tightly

  - There are appropriate signs warning other employees that this is a hazardous waste storage area

  - Drums are not stored onsite longer than you are allowed:
    - LQGs – 90 days
    - SQGs – 180 days or 270 days if the treatment, storage, and disposal facility is more than 200 miles away

  - There is no more waste onsite than is allowed for your generator status

  - Drums containing incompatible hazardous waste are stored separately or protected by a structure, such as a dike or berm.

  Some states may require that you keep a written record of these inspections. Any problems should be corrected immediately. If any corrections are made, they should be noted in a permanent record and kept on file for at least three years.

- If your facility has outdoor accumulation areas, and if you are storing ignitable or reactive wastes, make sure that containers of these wastes are stored at least 50 feet from the your facility's property line as this creates a protective buffer zone.

- NEVER store two or more wastes in the same container that could react to cause fires, leaks, or other releases.

Personnel Training

Proper waste handling can save your facility money in waste treatment and disposal and in lost time due to employee illness or accidents. You must train your employees on the procedures for properly handling hazardous waste, as well as on

> *Keep in mind that employees who are responding to releases of hazardous substances and hazardous waste are required to be trained under OSHA's Hazardous Waste Operations and Emergency Response (HAZWOPER) requirements (see 29 CFR 1910.120).*

emergency procedures (40 CFR 262.34(a)). For LQGs, the training must be formalized and be completed by employees within six months of accepting a job involving the handling of hazardous waste, and you are required to provide annual review of the initial training.

It is important to note that training you may be required to conduct by OSHA differs from hazardous waste management training. Make sure you provide both types of training to your employees.

Hazardous Waste Shipment Labeling and Placarding

When you prepare hazardous wastes for shipment, you must put the wastes in properly labeled containers that are appropriate for transportation according to the DOT regulations (40 CFR 262). Your transporter should be able to assist you. You must:

- Write the manifest document number on the drum label. A blank space intended for this purpose is provided on hazardous waste labels available from label distributors.

- Label all drums using the four-inch DOT warning labels (available from the waste hauler or a label distributor), which are marked with the proper DOT shipping name and number according to DOT requirements. Usually your hauler will do this for you.

- Make sure that (if required) the hauler displays the proper 10-3/4" DOT placard on all four sides of the truck that hauls your hazardous waste. Although the hauler usually provides these, you are responsible for making sure your hauler displays the appropriate placard.

If you need additional information, you may wish to consult the requirements for packaging and labeling hazardous wastes found in the DOT regulations. To find out what the requirements are for your specific waste, you should contact your state transportation agency, your hauler, or your waste disposal/treatment facility who can help you understand the DOT requirements. It may be helpful for you to create a shipping manual with guidance for packing, shipping, and disposal/recycling of all wastes leaving your facility.

> **REMEMBER: Just because you have shipped the hazardous waste off your site and it is no longer in your possession, your liability has not ended. You are potentially liable for cleanup costs under Superfund for any mismanagement of your hazardous waste. The manifest will help you track your waste during shipment and make sure it arrives at its proper final destination.**

Reporting/Recordkeeping for Hazardous Waste Management Practices

Your food processing facility is required to meet various reporting and recordkeeping requirements as part of your hazardous waste management obligations. These requirements are summarized below:

Uniform Hazardous Waste Manifest. The Uniform Hazardous Waste Manifest Form (EPA Form 8700-22) is a multi-copy shipping document that reports the contents of your shipment, the transport company used, and the treatment/disposal facility receiving the wastes (40 CFR 262.20).

The manifest form is designed so that shipments of hazardous waste can be tracked from their site of generation to their final destination, or from "cradle-to-grave." The hazardous waste generator, the transporter, and the treatment/disposal facility must each sign this document and keep a copy. The waste disposal/treatment facility also must send a copy back to you, so that you can be sure that your shipment was received.

A copy of the manifest is required to be kept on file at your facility for three years, or until a signed copy of the manifest is received from the waste disposal/treatment facility. The signed copy of the manifest is required to be kept on file for three years. If you do not receive a signed copy from the waste treatment/disposal facility within 30 days, it is a good idea for you to find out why and, if necessary, let the state or EPA know (see Exception Reports).

> **Tolling Agreements**
> *Note: Small Quantity Generators (SQGs) are not required to prepare a hazardous waste manifest if: (1) the waste is reclaimed under a contractual (tolling) agreement which specifies the type of waste and frequency of shipments, and the vehicle used to transport the waste to the recycling facility and to deliver regenerated material back to the generator is owned and operated by the reclaimer of the waste; and (2) the generator maintains a copy of the reclamation agreement in his files for a period of at least three years after termination or expiration of the agreement (40 CFR 262.20(e)).*

You can obtain blank copies of the manifest form from several sources. To determine the best source for you, use this system:

- If the state to which you are shipping your waste has its own manifest, use that manifest form (your waste transporter will know which manifest form is required). Contact the hazardous waste management agency of that state, your transporter, or the waste treatment/disposal facility to obtain manifest forms.

- If the state to which you are shipping your waste does not have its own manifest, use the manifest of the state in which your waste was generated. Contact your transporter or your state hazardous waste agency for blank forms.

- If neither state requires a state-specific manifest, you may use the "general" Uniform Hazardous Waste Manifest (EPA Form 8700-22). Copies are available from some haulers and waste treatment/disposal facilities.

When you sign the certification on ITEM 16 of the manifest form, you are personally confirming that:

- The manifest is complete and accurately describes the shipment.

- The shipment is ready for transport.

- You have reduced the amount and hazardous nature of your wastes to the greatest extent possible (within your budget constraints).

Transporters, recyclers, and waste treatment/disposal facilities may require additional information. Check with them before you prepare a hazardous waste shipment. States may also have additional requirements that must be followed. Your hazardous waste hauler often will be the best source for packaging and shipping information and will help in completing the manifest. If you have any trouble obtaining, filling out, or using the manifest, ask your hauler or your waste treatment/disposal facility operator.

Exception Reports. Exception reports document a missing return copy of the hazardous waste manifest. Reports must include a copy of the manifest and a cover letter signed by the generator or authorized representative explaining the efforts taken to locate the waste and the results of those efforts. If you are an SQG, you must submit a copy of the manifest with some indication that you have not received confirmation of delivery within 60 days of the date your waste was accepted for transport. If you are an LQG, you must contact the transporter and/or waste treatment/disposal facility within 35 days of the transporter accepting your waste for shipment to determine its status and submit a report within 45 days. Copies of exception reports must be maintained for three years.

Biennial Reports. If you are an LQG of hazardous waste, you must submit a biennial report (EPA 8700-13A) on March 1 of each even-numbered year to the appropriate state regulatory office (40 CFR 262.41). Some states impose this requirement on SQGs. Biennial report applications and instructions can be obtained from your state office. This report must contain the following information:

- Your EPA ID number, name, and address.

- The EPA ID number, name, and address of each offsite waste treatment/disposal facility where waste was sent during the year.

- A record of your hazardous waste activities for the previous calendar year, including the quantity of waste you generated or accumulated onsite and the quantity you sent offsite.

- A description of your efforts that year to reduce the toxicity and volume of hazardous wastes generated (i.e., waste minimization efforts).

- A description of the changes in volume and toxicity of waste actually achieved.

- The certification signed by you, the generator, or authorized representative.

Copies of biennial reports must be maintained onsite for three years.

Land Disposal Restriction Notification. Land disposal restrictions (LDRs) are regulations prohibiting the disposal of hazardous waste on land without prior treatment of the waste (40 CFR 268). The LDR notification ensures proper treatment of the waste prior to disposal. Recent changes in this regulation have decreased the reporting and recordkeeping burden for you (Federal Register, Volume 62, Number 91; May 12, 1997). Under this amended rule, you are required to provide a **one-time notification** about your waste to the treatment or disposal facility with the first shipment of waste offsite, and keep a copy in your files. The one-time notification applies to shipments of all restricted hazardous waste. No new notification is required unless there is a change in the waste, process, or receiving facility. A change in the waste is a change that affects the determination of which treatment standard applies. If this occurs or there is a change in the receiving facility, you must send a new notice to the receiving facility and place a copy of the new notice in your files (40 CFR 268.7).

The LDR notification must include:

- • EPA hazardous waste code for the wastes (e.g., F002).
- • Corresponding treatment standard(s) as identified in the federal RCRA regulations.
- • Manifest number for that shipment.
- • Certification statement.

As presented in Table 8-2, a LDR notification is required for SQGs and LQGs, unless a tolling agreement (see box titled *Tolling Agreements* on p. 8-16). Copies of the land disposal restriction notification form(s)and the waste analysis reports must be kept for three years (40 CFR 268.7(a)(8)). All records kept in connection with the LDR program may be stored electronically.

Summary of Recordkeeping Requirements. You are required by EPA to keep certain records on file to show that good housekeeping practices and monitoring are being performed at your food processing facility. EPA requires that records be kept on file at your facility for three years (40 CFR 262.40). These records include:

- Laboratory analyses and waste profile sheets for determining whether wastes generated by your facility are hazardous.

- Copies of all hazardous waste manifests, LDR notification(s), and exception reports.

- Copies of all Notification of Hazardous Activity forms submitted to and received from the state or EPA.

- Copies of all personnel training plans and documentation that indicate employees have completed the required training (LQGs only).

- • Copies of your facility's contingency plan (LQGs only).

- • Copies of your facility's biennial report (LQGs only).

It is a good idea to have these documents filed neatly in one place at your facility. State or federal inspectors will likely ask for copies of these documents while inspecting your facility.

Contingency Planning and Notification

A contingency plan will help you look ahead and prepare for accidents that could possibly occur at your food processing facility (40 CFR 262). If you are an LQG, you are required to have a **written contingency plan**. If you are an SQG, you must have **basic contingency procedures** in place. Although a **written** contingency plan is not federally required for SQGs or CESQGs, it is strongly recommended. It is also important to check with your state and local authorities for any additional contingency plan or emergency preparedness requirements. Table 8-3 presents the contingency plan requirements for LQGs and SQGs.

A contingency plan can be thought of as a set of answers to a series of "what if" questions. For example, "What if there is a fire in the area where solvents are stored?" or "What if I have a spill of hazardous waste or one of my containers leaks?" Emergency procedures are the steps that you should follow if you have an emergency. It is a good idea to make a list of these "what if" questions and to write down specific steps that you would take if the emergency occurred. Review these with your employees so they are also informed about their responsibilities in the event of an emergency.

Emergency Procedures and Accident Prevention

Emergency Planning. Your contingency plan, discussed above, must contain facility-specific details on what you have to do if you have an emergency. Specifically:

- • In the event of a **fire, explosion, or accidental release** of hazardous waste, you must immediately notify the National Response Center if the fire, explosion, or other release could threaten human health outside your food processing facility or when the release has reached surface water.

> *If you store oil onsite, you may be subject to specific prevention and response planning requirements. See Section 4.6 How Do I Comply With Oil Pollution Prevention Regulations? for more information.*

Table 8-3. Contingency Plan Requirements for LQGs and SQGs

LQG Contingency Plan Contents	SQG Contingency Procedures
Written plan required. he contingency plan must contain: • Instructions on what to do immediately whenever there is a fire, explosion, or release. • The arrangements agreed to by local police and fire departments, hospitals, and state and local emergency response teams to provide emergency services. • The names, addresses, and phone numbers of all persons qualified to act as emergency coordinator. • All emergency equipment at the facility. • An evacuation plan. Copies of the contingency plan must be submitted to the local police and fire departments, hospitals, and state and local emergency response teams that may be called upon to provide emergency services. You should maintain documentation showing that local authorities have been notified.	Basic plan required (not required to be written). The contingency procedures include the following: • You must have an emergency coordinator (employee) either at the facility or on call who is responsible for coordinating all emergency response measures. • You must post next to the telephone: (1) the name and number of the emergency coordinator; (2) the locations of the fire extinguishers and spill control material; and (3) the telephone number of the fire department. • You must ensure that all employees are thoroughly familiar with proper waste handling and emergency procedures.

The Center operates a 24-hour toll free number: **1-800-424-8802**, *or in Washington, DC:* **1-202-426-2675**. As soon as possible, you should have the hazardous waste and any contaminated materials or soils cleaned up by an appropriately trained person. Reporting any release and threat of release is required of SQGs and LQGs.

**IMMEDIATELY CALL
THE NATIONAL RESPONSE CENTER (1-800-424-8802) if:**

- If you have a serious emergency,
- You have placed a call to your local fire department, or
- You have a spill that extends outside of your facility boundaries or a spill that could reach surface water.

GIVE THEM THE INFORMATION THEY ASK FOR.
If you did not need to call, they will tell you.

ANYONE WHO IS REQUIRED TO CALL AND DOES NOT IS SUBJECT TO A $10,000 FINE, A YEAR IN JAIL, OR BOTH. If you are an owner or manager of a food processing facility and you fail to report a hazardous waste release, you may have to pay for the entire cost of repairing any damage, even if your facility was not the single or main cause of the damage.

During your telephone call to the National Response Center, you must give the following information:

- • Your facility name, address, and EPA identification number
- • The date, time, and type of incident (e.g., spill, fire)
- • The quantity and type of hazardous waste involved in the incident
- • The extent of injuries
- • An estimate of the quantity and location of any recovered materials, if any.

In addition, reports from LQGs must include an assessment of actual or potential hazards to human health or the environment (where this is applicable).

As stated above, the RCRA regulations require that emergency phone numbers and locations of emergency equipment must be posted near telephones. This means that **next to the phone** you must post:

- • The name, office phone number, home phone number, and address of your emergency coordinator.

- • A site plan or list of nearby:
  – Portable fire extinguishers
  – Special extinguishing equipment (e.g., foam, dry chemicals)
  – Fire alarms (only if not directly connected to fire department)
  – Spill control equipment (e.g., absorbent cotton rags)
  – Decontamination equipment (e.g., safety shower, eye wash fountain)
  – Water at adequate volume and pressure (e.g., water hoses, automatic sprinklers, water spray systems).

- • The telephone numbers of the fire and police departments.

- • Although not required, it is strongly recommended that you also post the following phone numbers by the telephone:
  – State or local emergency response teams
  – Hospital
  – Local ambulance service
  – National Response Center
  – State Department of Public Safety.

All employees who deal with hazardous waste must know proper waste handling and emergency procedures. An employee must be appointed to act as the emergency coordinator to ensure that emergency procedures are carried out in the event of an emergency.

> *Keep in mind that employees who are responding to releases of hazardous substances and hazardous waste are required to be trained under OSHA's Hazardous Waste Operations and Emergency Response (HAZWOPER) requirements (see 29 CFR 1910.120).*

The emergency coordinator (or someone designated by that person) must:

- • Be available 24 hours a day either at the facility or by phone
- • Know whom to call and what steps to follow in an emergency
- • Commit company resources as necessary to respond to an emergency.

Because most food processors are small businesses, the owner or operator probably already performs these functions. Therefore, it is not intended (nor is it likely) that you will need to hire a new employee to fill this role.

Accident Prevention. In accordance with RCRA, your facility must have appropriate cleanup materials and emergency communication equipment for handling hazardous waste at your site. Some of the steps you may need to take to prepare for emergencies at your facility include the following:

- Make sure that there are no floor drains near the area where solvents are used that lead to the sewer, septic tank, or storm water drain.

- Store hazardous waste in areas away from doorways. The floor in your storage area should be leak-proof (e.g., concrete with an epoxy coating). If there is a doorway nearby, a concrete barrier is required to prevent the flow of material out of the door in case of a large spill.

- Provide room for emergency equipment and response teams to get into any area in your facility in the event of an emergency.

- If you are an LQG, you must write to local fire, police, and hospital officials or state or local emergency response teams explaining that you handle hazardous wastes and ask them for their cooperation and assistance in handling emergency situations.

- Install and maintain emergency equipment (e.g., an alarm, a telephone, two-way portable radios, fire extinguishers, hoses, and automatic sprinklers) at your facility in hazardous waste storage areas so that it is immediately available to your employees if there is an emergency. This equipment should be inspected monthly.

8.7   Underground Storage Tanks (USTs)

According to EPA, an underground storage tank (UST) is "any one or combination of tanks (including underground pipes connected thereto) that is used to contain an accumulation of regulated substances, and the volume of which (including the volume of underground pipes connected thereto) is ten percent or more beneath the surface of the ground."

> *For information on aboveground storage tank (AST) requirements, see Section 4.6 How Do I Comply With Oil Pollution Prevention Regulations?*

USTs are subject to strict state and federal requirements. Federal regulations of USTs, contained in 40 CFR 280, require that all regulated UST systems be designed and constructed to retain their structural integrity throughout their operating life, and all USTs and attached piping be protected from corrosion. In addition, all systems must be equipped with spill and overfill protection and leak detection monitoring.

Subtitle I of RCRA governs activities and requirements related to UST systems. Subtitle I established a new and comprehensive regulatory program for UST systems containing petroleum products or substances defined as hazardous under CERCLA. Subtitle I includes the following provisions for UST systems:

- • Design, construction, installation, operating, and notification requirements for new and existing systems;
- • Release detection, reporting, investigation, confirmation, release response, and corrective action for systems containing petroleum or hazardous substances; and
- • System closure requirements.

States generally have the same requirements as RCRA Subtitle I. However, some states (and municipalities) have more stringent UST regulations. You should contact your state UST office and your local municipality to determine if there are additional UST regulations with which you must comply.

The federal UST regulations do not apply to:

- • Tanks with a capacity of 110 gallons or less;
- • Farm or residential tanks holding 1,100 gallons or less of motor fuel used for noncommercial purposes;
- • Tanks storing heating oil, where the oil is used on the property where it is stored;
- • Tanks on or above the floor in underground areas (e.g., basements);
- • Septic tanks and systems for collecting storm water and wastewater;
- • Flow-through process tanks; and
- • Emergency spill and overfill tanks.

> *Note: EPA defines "heating oil" to include fuel for boilers for process steam generation (40 CFR 280.12).*

Requirements for notification, recordkeeping, leak detection, and spill, overfill, and corrosion protection are described below. Federal and state laws mandate strict penalties for failure to report or to respond properly to spills or leakage once detected. Penalties also apply to violations of the requirements for the installation, monitoring, testing, registration, and removal or closure of USTs.

Notification. A facility must report to the regulatory authority on the following occasions:

- • **UST installation**. When an UST is installed, fill out a notification form. The form must be submitted to the responsible state UST program office for all onsite UST systems. The Notification Form includes certification of compliance with federal requirements for installation, cathodic protection, release detection, and financial responsibility for UST systems installed after December 22, 1988.

·· **Suspected release**. Report suspected releases to the regulatory authority. If a release is confirmed, the facility must also report follow-up actions planned and take corrective actions to correct the damage caused by the UST.

· **UST closure**. Notify the regulatory authority 30 days before the facility permanently closes the UST.

Recordkeeping. Leak detection, corrosion protection, financial responsibility, closure, and corrective action records must be maintained onsite.

· Leak detection records include: the past year's monitoring results and most recent tightness test; copies of performance claims; and maintenance, repair, and calibration of leak detection equipment.

· Corrosion protection records include results of the last two tests proving the cathodic protection system is working and the last three inspections proving that impressed current systems are operating properly.

· If you are an owner and operator of a petroleum UST (e.g., vehicle fuel tank), you must have financial responsibility documentation showing you (1) have either insurance coverage; a guarantee from another firm; a surety bond; or a letter of credit; (2) have passed a financial test; (3) have a trust fund; (4) participate in a state financial assurance fund; or (5) use another financial method(s) of coverage approved by your state. (Note: if the owner and operator of a petroleum UST are separate persons, only one person is required to demonstrate financial responsibility; however, both parties are liable in event of noncompliance.) (See 40 CFR 280.090.)

· Closure records document that the UST was removed from service in accordance with federal requirements for notification and correct, safe closure.

· Corrective action records document that any releases from USTs have been reported to the appropriate agency and have been responded to as required.

Leak Detection. Facilities must check their USTs at least once a month to see if they are leaking. The facility must conduct one of the three following methods of leak detection:

· All USTs can use **monthly monitoring** consisting of one of the following methods or other methods approved by the regulatory agency:

– Automatic tank gauging
– Vapor monitoring
– Interstitial monitoring with secondary containment
– Groundwater monitoring
– Statistical inventory reconciliation.

Check with the state UST program to determine which methods are acceptable.

•• USTs installed before December 22, 1998 can temporarily use monthly inventory control and annual tank tightness testing. **This is not an acceptable method after December 22, 1998.**

•• USTs installed or upgraded with spill, overfill, and corrosion protection can temporarily use monthly inventory control combined with tank tightness testing every five years. This combined method can be used only for ten years after the tank has been installed or retrofitted with corrosion protection or until December 22, 1998, whichever is later.

In addition, facilities must conduct leak detection on any pressurized piping by (1) monthly monitoring (as described above) or annual line testing, and (2) an automatic flow restrictor, an automatic shutoff device, or a continuous alarm system.

Spill, Overfill, and Corrosion Protection. USTs installed on or before December 22, 1988, must meet spill, overfill, and corrosion protection requirements no later than December 22, 1998. USTs installed after December 22, 1988 were required to be constructed with spill, overfill, and corrosion protection.

•• Spill and overfill protection. USTs must have catchment basins to contain spills. In addition, the facility and the fuel deliverer must follow industry standards for correct filling practices. New USTs must have overfill protection devices when they are installed. The three main types of overfill protection devices are automatic shutoff devices, overfill alarms, and ball float valves.

•• Corrosion protection. Corrosion is the dissolution or gradual wearing away of materials, especially by chemical action. Metal is especially susceptible to corrosion. If your UST or piping is made of metal or has metal components, you must have some form of corrosion protection for them. All corrosion protection systems must be operated and maintained to provide continuous corrosion protection to the metal components of the portion of the UST and piping that are in contact with the ground and routinely contain petroleum products or hazardous substances.

To find out more about federal UST requirements, you can receive free explanatory publications and assistance by calling EPA's RCRA/UST, Superfund and EPCRA hotline at 1-800-424-9346 or 703-412-9810, or by visiting EPA's Office of Underground Storage Tanks website at **http://www.epa.gov/OUST/**. State and local UST requirements can differ from federal requirements so be sure to check with appropriate state and local regulatory agencies.

8.8   Used Oil Management Standards

If you generate used oil at your food processing facility, you are responsible for ensuring that it is managed properly. A generator of used oil is defined by EPA as any business which produces used oil through commercial and industrial operations, or that collects it from

> ***What is used oil?*** *EPA's defines used oil as "any oil that has been refined from crude oil, or any synthetic oil, that has been used and as a result of such use is contaminated by physical or chemical impurities."*

these operations. The definition of used oil (see box) includes oils that are used as hydraulic fluids as well as oils that are used to lubricate automobiles and other machinery, cool engines, or suspend materials in industrial processes.

As a generator of used oil, you must follow the used oil management standards found in 40 CFR 279. Some of these requirements include:

- • Keeping storage tanks and containers in good condition

- • Labeling storage tanks and containers "Used Oil"

- • Cleaning up any used oil spills or leaks to the environment (see Section 9.0)

- • Using a transporter with an EPA ID number when shipping used oil offsite.

State Requirements

Used oil is not categorized as hazardous waste under federal RCRA requirements, however, it may be under some state regulations. For example, California classifies used oil as hazardous waste. In addition, used oil may have strict disposal requirements in some states. **Be sure to contact your state regulatory agency for information on how to manage used oil.**

8.9   Good Environmental Management Practices

8.9.1   How to Select a Hazardous Waste Transporter and Waste Disposal/Treatment Facility

Carefully choosing a transporter and designating a waste treatment/disposal facility are important. Even when the waste leaves your control, **your facility remains legally responsible** for the proper disposal of your waste and any associated spills or accidents.

You should be aware that under RCRA, transporters are required to:

- Obtain an EPA/state identification number
- Comply with the manifest system
- Respond appropriately to hazardous waste discharges
- Comply with both the RCRA requirements (40 CFR 263) and the DOT regulations (49 CFR 171-179).

Before choosing a transporter or designating a facility, check with the following sources:

- Your state hazardous waste management agency or EPA regional office, which will be able to tell you whether or not a company has an EPA/state identification number and may know whether or not a company has had any problems. They may also have a list of licensed (approved) transporters.

- Your friends and colleagues in the food processing business who may have used a specific hazardous waste transporter or treatment/disposal facility in the past.

- Your trade association(s) which may keep a file on companies that handle hazardous waste.

- Your Better Business Bureau or Chamber of Commerce to find out if any complaints have been registered against a transporter or facility.

After checking with these sources, contact the transporter and hazardous waste treatment/disposal facility directly to verify that they have an EPA/state identification number, and that they can and will handle your waste. In some states, the transporter and the designated facility may be required to have a special permit to operate. Make sure that the transporter and waste treatment/disposal facility have the necessary permits and insurance and that the transporter's vehicles are in good condition. You may also want to ask them:

- Where the waste is going
- To provide information on their track record
- If they have ever been cited for improper practice.

Checking sources and choosing a transporter and waste treatment/disposal facility may take some time. You should begin checking before you open your shop or well ahead of the time you will need to ship your waste.

8.9.2   Disposing of Hazardous Waste Onsite

You may **not** dispose of your hazardous waste on your property. Additionally, if you discharge more than 15 kg of a *hazardous substance* in a month to the POTW, you must meet certain reporting requirements (see Table 4.7 *Reporting Requirements for All Indirect Discharges*). Typically, discharging any hazardous substance to a sewer is not considered good management practice and in many states it may be illegal. For more information, contact the local wastewater or sewage treatment office or your state hazardous waste management agency.

8.9.3   Good Housekeeping

Good hazardous waste management can be thought of as simply "good housekeeping" practices which include:

- Using fewer hazardous materials
- Reusing materials as much as possible

- Recycling and reclaiming waste
- Reducing the amount of waste you generate.

To reduce the amount of waste you generate:

- Do not mix nonhazardous wastes with hazardous wastes (e.g., combining nonhazardous cleaning agents or rags in the same container as hazardous wastes). If you do, the nonhazardous wastes become subject to hazardous waste regulations and you will have more hazardous waste for disposal.

- Avoid mixing several different hazardous wastes. Doing so may make recycling very difficult, if not impossible. It may also make disposal more expensive.

- Avoid spills or leaks of hazardous products. The materials used to clean up such spills or leaks also become hazardous wastes.

- Make sure the original containers of hazardous products are completely empty before you throw them away. Use ALL of the product.

- Avoid using more of a hazardous product than you need. For example, use no more cleaning solvent than you need to do the job.

Reducing your hazardous waste means saving money on raw materials and reducing the costs to your business for managing and disposing of your hazardous wastes.

SECTION 9 CONTENTS

9. HOW DO I COMPLY WITH SPILL OR CHEMICAL RELEASE REQUIREMENTS?

9.1 Introduction

This section provides an overview of the emergency planning, reporting, notification, and response requirements across major statutes administered by EPA, (plus related OSHA requirements) that apply to: (1) the storage of oils and hazardous materials and (2) the spills and releases of pollutants, such as oils, hazardous materials and wastes. The term "hazardous material" is used here in the same way as used in Section 3.0 *Understanding the Process: Inputs, Outputs and Applicable Federal Environmental Regulations.* Because these requirements are explained in detail in the statue-

> ***Check State and Local Requirements:*** *Your state and municipality may have requirements in addition to the federal requirements with which you must comply. You should check with your state or local environmental agencies to determine these requirements.*

specific sections of the guide, this section does not repeat this information, but rather, briefly compares and contrasts the regulatory approaches and the different terminology for regulated substances.

The primary purpose of this section is to assist you in understanding the similarities and differences among these requirements and their interconnections, and to provide a single place of reference against which you should check the applicable regulatory requirements for your inputs and waste outputs. Another purpose is to re-emphasize the importance of complying with these requirements, some of which are relatively new, such as the Risk Management Program under the Clean Air Act. Please read this section very carefully, become knowledgeable about the relationship between these requirements, and make sure that your staff knows what to do in the event of an emergency.

Regulatory Approaches

Some federal statutes, such as the Clean Air Act (CAA), Clean Water Act (CWA), Oil Pollution Act (OPA), and the Resource Conservation and Recovery Act (RCRA), have planning, reporting, notification and response requirements that pertain to spills or releases to **specific environmental compartments** (e.g., air, water, land). For example:

- CWA and OPA requirements pertain to spills or releases to surface water and/or groundwater.

- CAA requirements apply to releases to ambient air.

- RCRA requirements pertain to spills or releases to land.

In addition, some requirements under the CWA, OPA, and RCRA are designed to alert publicly owned treatment works (POTWs) of impacts from spills of oil(s), hazardous waste, or other hazardous materials into the sewer system. In addition, there are requirements to notify appropriate authorities of discharges of hazardous waste to septic systems.

In contrast to this approach, the planning, reporting, notification, and response requirements under the Emergency Planning and Community Right-to-Know (EPCRA) and the Comprehensive Environmental Response, Compensation, and Liability Act (CERCLA) focus on **specific substances**, irrespective of the environmental compartment into which a spill or release might occur.

For either approach, the particular substances subject to the requirements are defined through the regulations under each federal statute. The terms used for regulated materials under each program differ, and this can be confusing. Table 9-1 *Terms for Regulated Materials Under Various Statutes* shows terms used under the six (6) statutes administered by EPA, plus OSHA. Each program defines and usually lists the materials or substances that are regulated. A material or substance may appear on one or more of these lists. There is, however, no single "master" list to which you can refer. Also these lists may be modified or amended from time to time. Therefore, you should obtain and review each definition and list to determine what materials or substances at your facility may be regulated, and you should keep abreast of relevant changes.

Table 9-1. Terms for Regulated Materials Under Various Statutes

Statute	Term Used	CFR citation
EPCRA	Extremely hazardous substances	40 CFR 355
OSHA	Hazardous chemicals	29 CFR 1910
CERCLA	Hazardous substances	40 CFR 302
CWA	Hazardous substances	40 CFR 116 and 117
OPA	Oil (petroleum or non-petroleum)	40 CFR t 112
RCRA	Hazardous wastes	40 CFR 261
CAA	Regulated substances	40 CFR 68

Organization of this Section

Planning and reporting activities are critical in preventing accidental releases of regulated materials at any facility. For the purposes of this guide, Section 9.2 *Emergency Planning and Reporting Requirements* groups these requirements across several statutes. The term "reporting" is used in a particular way as highlighted in the text box. Similarly, Section 9.3 *Notification and Response Requirements* groups the notification and response requirements across several statutes This structure for dividing and grouping these requirements was selected solely for the purpose of comparing and contrasting similar requirements across statutes. Keep in mind, however, that implementation of these requirements– for planning, reporting, notification, and response to releases–at your facility should be integrated.

> The term "reporting" as described in Section 9.2 Emergency Planning and Reporting Requirements, refers to the inventory reports that must be made regarding the hazardous chemicals that are stored onsite at a facility. This type of inventory reporting should take place before a spill occurs to ensure proper emergency response to the spill. This reporting is different from the notification requirements (described in Section 9.3 Notification and Response Requirements) triggered when accidental spills or releases do occur.

9.2 Emergency Planning and Reporting Requirements

This section provides an overview of the planning and reporting requirements under each statute. This will allow you to see the similarities and differences, and the interconnections among these requirements. Table 9-2 *Major Federal Regulations With Planning and Reporting Requirements* provides references to these requirements under EPA's statutes, plus OSHA. For a fuller explanation of these planning requirements, refer to the statute-specific sections of the guide for information and references.

Table 9-2. Major Federal Regulations With Planning and Reporting Requirements

Subject	Law	Reference to Regulation
EPCRA Planning Requirements	EPCRA	40 CFR 355
EPCRA Hazardous Chemical Inventory	EPCRA	40 CFR 370
EPCRA Toxic Chemical Release Reporting	EPCRA	40 CFR 372
SPCC Plans	CWA and OPA	40 CFR 112
OPA Facility Response Plans	CWA amended by OPA	40 CFR 112
CAA Risk Management Plans	CAA Section 112(r)	40 CFR 68
RCRA Contingency Plans	RCRA	40 CFR 262
OSHA Process Safety Management[1]	OSHA	29 CFR 1910
Hazard Communication Standard[1]	OSHA	29 CFR 1910
Hazardous Waste Operations and Emergency Response (HAZWOPER)[1]	OSHA	29 CFR 1910

[1] These requirements are mentioned here because of their relationship with EPCRA, however, they will not be discussed in detail. Please see 29 CFR 1910 for more information.

Table 9-2 includes OSHA requirements because of the relationship between several OSHA programs and EPA's programs--for example, the relationship between CAA Risk Management Plans and OSHA Process Safety Management, and the relationship between EPCRA Hazardous Chemical Inventory Reporting and OSHA's Hazard Communications Standard.

Emergency planning and reporting are important activities at every facility. Your facility can prepare for an accidental spill or release by creating required prevention and response plans, and by participating in local emergency planning activities. A new approach to handle planning requirements is to combine two or more plans into one integrated plan containing all the requirements for your facility. **Before doing this, you should check with your state or your EPA Regional Office to find out if this is an accepted practice.**

9.2.1 EPCRA Emergency Planning and Reporting– Other Than Section 313

Section 7.0 *How Do I Comply With the Emergency Planning and Community Right-to-Know Act Regulations?* is the primary section in this guide to refer to for information about EPCRA requirements. This section contains only a very brief summary. Keep in mind the following, critical distinctions among the sections of EPCRA, as presented in Table 9-3. First, Sections 301-303, 311, and 312 focus on the amount of chemicals **present at** your facility. Whereas Section 313 focuses on the amounts of chemical **manufactured, processed, or otherwise used**.

Table 9-3. Summary of EPCRA Regulatory Criteria

Subject	Statutory Section	Regulatory Criteria	Reference to Regulation
Emergency Planning	Sections 301-303	"Present at"	40 CFR 355
Hazardous chemical inventory and reporting	Sections 311 and 312	"Present	40 CFR 370
Toxic chemical release reporting	Section 313	"Manufactured, processed or used"	40 CFR 372

Emergency Planning. Congress designed the emergency planning sections (Sections 301-303) of EPCRA to develop state and local governments' emergency response and preparedness capabilities through better coordination and planning, especially with the local community. The responsibility of your facility to participate in emergency planning activities depends on the presence of EPCRA extremely hazardous substances (EHSs) in amounts equal to or in excess of the threshold planning quantities (TPQs). EHSs typically found at food processing facilities include ammonia (for refrigeration), chlorine (for disinfection), and nitric and sulfuric (for cleaning) acids. See Section 7.2 *Emergency Planning* for more information about these requirements.

Hazardous Chemical Inventory and Reporting (MSDS and Tier Reporting). The hazardous chemical inventory and reporting provisions outlined in EPCRA Sections 311 and 312 require you to take an inventory of the EPCRA EHSs (40 CFR 355, Appendices A and B) and the OSHA hazardous chemicals present onsite at your facility in amounts equal to or in excess of TPQs. Your facility is subject to these reporting requirements if: (1) your facility is regulated under OSHA's Hazard Communications Standard (29 CFR 1910.1210), and (2) your facility equals or exceeds EPA-established thresholds for EPCRA EHSs and OSHA hazardous chemicals onsite at any one time. The thresholds that you must use varies depending on whether the chemical is classified as an EPCRA EHS or OSHA hazardous chemical. See Section 7.4 *Hazardous Chemical Inventory and Reporting* for more information on these requirements.

Exemptions from these requirements exist for specific chemicals under OSHA and EPCRA Section 311(e). One exemption under Section 311(e) covers any food, food additive, color additive, drug or cosmetic regulated by Food and Drug Administration (FDA). Remember, these exemptions apply to specific chemicals within the scope of the exemption only, not to all hazardous chemicals at a particular facility.

- ***MSDS and Hazardous Chemical Inventory Reporting.*** Under Section 311, you must submit a **one-time notification** of the EPCRA EHSs and OSHA hazardous chemicals present at your facility in amounts equal to or in excess of the TPQs to the SERC, LEPC, and local fire department. To meet the reporting requirement, your facility must submit either an MSDS **or** a list of the EPCRA EHSs and OSHA hazardous chemicals grouped by hazard category. Information reported under Section 311 must be updated.

-- ***Tier Reporting.*** Under Section 312, your facility must meet an **annual reporting** requirements for EPCRA EHSs and OSHA hazardous chemicals in amounts equal to or in excess of threshold levels. If your facility equals or exceeds the threshold levels **at any time** (not the average amount onsite) in the preceding year, you must submit to the SERC, LEPC, and local fire department an "Emergency and Hazardous Chemical Inventory Form." This form must be submitted by March 1 and it covers the previous calendar year.

  EPA publishes two types of inventory forms, **Tier I** and **Tier II**, for reporting inventory information. While federal regulations require only the submission on a Tier I form, EPA encourages, and some states require, the use of the Tier II form. EPA distributes an electronic version of the Tier II form in both Windows and DOS formats.

For more information on EPCRA planning and reporting requirements, refer to Section 7.0 in this Guide or contact the RCRA/UST, Superfund and EPCRA Hotline at 1-800-424-9346 or 703-412-9810, or access EPA's Chemical Emergency Preparedness and Prevention Office (CEPPO) homepage at **http://www.epa.gov/ceppo/**.

9.2.2 EPCRA Toxic Chemical Release Reporting– Section 313

Section 7.5 *Toxic Chemical Release Reporting - EPCRA Section 313* is the primary section of this guide to refer to for information about EPCRA Section 313 reporting requirements. To assist food processing facilities in complying with the reporting requirements of EPCRA Section 313 and Section 6607 of the Pollution Prevention Act of 1990, EPA's Office of Pollution Prevention and Toxics (OPPT) has prepared a guidance manual, entitled *EPCRA Section 313 Reporting for Food Processors* (EPA 745-R-98-011, September 1998). This new guidance supplements the *TRI Forms and Instructions*. For more information, visit the TRI Homepage, **http://www/epa/gov/opptintr/tri.**)

EPCRA Section 313 requires certain designated businesses to submit annual reports to EPA (commonly referred to as Form R and Forms A) on the amounts of more than 600 EPCRA Section 313 chemicals and chemical categories released and otherwise managed (40 CFR 372). The standard report is Form R. However, to reduce the reporting burden for small businesses, EPA established an alternative threshold reporting level, known as Form A.

EPCRA Reporting Criteria. The following four questions will help you to determine if your facility must prepare a Form R report.

1. Is the SIC Code for your facility included in the list covered by EPCRA Section 313 reporting?

2. Does your facility employ 10 or more full time employees or their equivalent?

3. Does your facility manufacture (which includes importation), process, or otherwise use EPCRA Section 313 chemicals?

4. Does your facility exceed any applicable thresholds of EPCRA Section 313 chemicals?

If you answer "**No**" to <u>any one</u> of the first three questions, you are not required to prepare and submit Form R or Form A reports. If you answer "**Yes**" to <u>all</u> of the first three questions, you must then address question four.

Threshold Determinations. To address question four, you must do the following: a) complete a threshold calculation for each EPCRA Section 313 manufactured, processed or otherwise used at your facility; b) compare these calculations to the reporting thresholds shown below in Table 9-4; and c) for each EPCRA section 313 chemical exceeding a threshold, you must submit a Form R (or a Form A) report.

EPCRA Section 313 reporting requirements define three activity categories for each EPCRA Section 313 chemical. These categories include "manufacturing" (which includes importing), "processing" or "otherwise using."

Table 9-4. EPCRA Section 313 Activity Categories/Reporting Thresholds

Chemical Activity	Activity Threshold
Manufacturing	25,000 pounds/year
Processing	25,000 pounds/year
Otherwise Use	10,000 pounds/year

These activity thresholds apply to each EPCRA Section 313 chemical. Because each category is mutually exclusive of the others, you must conduct a separate threshold determination for each chemical for each activity category. The threshold determination is based **solely** on the quantity **actually** manufactured (including imported), processed, or otherwise used, **not** on the quantity of chemicals stored on-site or purchased.

Form R/Total Annual Reportable Amount. If your facility exceeds any one of the activity thresholds, then you must submit a Form R (provided that you also meet the SIC code and employee criteria). Among the information you must report for each EPCRA 313 chemical or chemical category that exceeds a reporting threshold, is the "**total annual reportable amount**" of "**releases and other waste management activities.**" This amount is defined as follows: the sum of the on-site amounts released (including disposal), treated, combusted for energy recovery, and recycled, combined with the sum of the amounts transferred off-site for recycling, energy recovery, treatment, and release (including disposal). This total corresponds to the total of data elements 8.1 through 8.7 on the 1997 version of the Form R.

Form A. The Form A report, also referred to as the "Certification Statement "(59 FR 61488, November 1994), is an alternative to Form R. EPA developed Form A to reduce the annual reporting burden for facilities that meet certain criteria. The Form A Certification Statement must be submitted for each eligible EPCRA Section 313 chemical. Form A does **not** require your facility to report any estimate of releases and/or other waste management activities. Rather, your facility must simply certify that the total annual reportable amount does **not exceed 500 pounds** for that particular chemical.

EPCRA Section 313 Recordkeeping. EPA requires you to maintain records substantiating the Form R or Form A submission, for a minimum of three years (40 CFR 372.10). Each facility must keep copies of the Form R or Form A along with all supporting documents, calculations, work sheets, and other forms that you use to prepare the Form R or Form A. EPA may request this supporting documentation during a regulatory audit.

9.2.3 Oil Spill Prevention Plans (SPCC) and Response Plans (FRPs)

SPCC Plans. EPA first issued the Oil Pollution Prevention Regulation, known as the Spill Prevention, Control and Countermeasures (SPCC) regulation, under the CWA in 1973. It established requirements, including the SPCC Plan, for facilities to **prevent** oil spills from reaching navigable waters of the U.S., or adjoining shorelines (40 CFR 112.3 through 112.7). These requirements apply to **nontransportation-related facilities** that meet these criteria:

- Could reasonably be expected to discharge oil in harmful quantities into navigable waters of the United States or upon the adjoining shorelines, **AND**

- Have (1) an **aboveground** oil storage capacity of more than 660 gallons in a single container; **or** (2) a total aboveground oil storage capacity of more than 1,320 gallons; **or** (3) a total **underground** storage capacity of more than 42,000 gallons.

If your facility is subject to the SPCC requirements based on the above description, EPA requires you to prepare an SPCC plan and conduct an initial screening to determine whether you are required to develop a Facility Response Plan (FRP) (see below). Those facilities that could cause **substantial harm** to the environment must prepare and submit an FRP to EPA for review. **Significant and substantial harm** facilities also must meet these requirements.

> *SPCC-regulated facilities must also comply with other federal, state, or local laws, some of which may be more stringent.*

The SPCC Plan is a written site-specific description detailing how a facility's operation complies with 40 CFR 112. In order to comply with 40 CFR 112, the SPCC Plan must be a carefully thought out plan, prepared in accordance with good engineering practices and which has the approval at a level with the authority to commit the necessary resources. While each SPCC Plan is unique, certain elements must be included in order for the SPCC Plan to comply with the provisions of 40 CFR 112. If a section does not apply to your facility, your Plan must state this. See Section 4.6.2 *SPCC Requirements* for specific elements to include in your SPCC plan.

Facility Response Plans. In 1990, Congress passed the Oil Pollution Act (OPA) which amended Section 311 of the Clean Water Act to require **substantial harm** facilities to develop and implement FRPs. EPA's FRP requirements appeared as a final rule in the Federal Register on July 1, 1994 (40 CFR 112.20 and 112.21 and include Appendices A through F). Under the FRP requirements, owners and operators of facilities that could cause "substantial harm" to the environment by discharging oil into navigable water bodies or adjoining shorelines must prepare plans for responding, to the maximum extent practicable, to the worst case discharge and to a substantial threat of such a discharge of oil.

Facilities subject to the FRP requirements under 40 CFR 112.20 are referred to either as **substantial harm** facilities or **significant and substantial harm** facilities. FRPs from substantial harm facilities are **reviewed** by EPA while FRPs from significant and substantial harm facilities are **reviewed and must be approved** by EPA.

If subject to the FRP requirements, you must prepare and submit a FRP to the appropriate EPA Regional Office. To assist you in preparing a FRP, EPA has prepared and included a **model facility response plan** in 40 CFR 112.2, Appendix F. EPA recognizes that there may be many facilities with existing response plans prepared to meet other requirements. Under OPA, you do not need to prepare a separate FRP provided that the original response plan:

(1) satisfies the appropriate requirements and is equally as stringent;
(2) includes all elements described in the model plan;
(3) is cross-referenced appropriately; and
(4) contains an action plan for use during a discharge.

EPA also recognizes that many facilities have established SPCC plans. Although response plans and prevention plans are different, and should be maintained separately, some sections of the plans may be the same. Under OPA regulations, you are allowed to reproduce or use those sections of the SPCC plan in the FRP. For more information on FRPs, see Section 4.6.3 *Facility Response Plans* or visit EPA's Oil Program Homepage at **http://www.epa.gov/oilspill/**.

9.2.4 CAA Risk Management Planning

As required under Section 112(r) of the amended CAA, EPA has promulgated the Risk Management Program Rule (40 CFR 68). The rule's main goals are to **prevent** accidental releases of regulated substances and to **reduce** the severity of those releases that do occur by requiring facilities to develop risk management programs. A risk management program must incorporate three elements: a hazard assessment, a prevention program, and an emergency response program. These elements are to be summarized in a risk management plan that will be made available to state and local government agencies and the public. Besides helping facilities prevent accidents, the rule can improve the efficiency of work operations by ensuring that workers are trained in proper procedures and by using preventive maintenance to reduce equipment breakdowns.

If your facility has more than a **threshold** quantity of any of the **140 regulated substances** in a single process, you are required to develop a risk management program and to summarize your program in a risk management plan by **June 21, 1999**. The plan you submit to EPA must summarize your program and must be made available to the public. Once your plan is submitted, it will be reviewed for accuracy and completeness. A site visit also may be conducted at your facility by either EPA, state, or local officials to determine whether your plan accurately reflects your risk management program in operation.

> **Regulated Substances.**
> EPA listed 140 regulated substances by rule published January 31, 1994; and amended the list by rule on December 18, 1997. EPA may amend the list in the future as needed.

To make compliance easier for small businesses, EPA has worked with trade associations and other industry groups to develop a series of **industry-specific brochures** that will assist businesses in creating their risk management programs. In addition, EPA has been working with industry groups (e.g., ammonia refrigeration) to develop model risk management programs. To review this **model program**, refer to EPA's Chemical Accident Prevention and Risk Management Planning website at **http://www.epa.gov/swercepp/acc-pre.htm#Model Plans/**.

For more information about risk management planning requirements and industry-specific brochures, see EPA's Chemical Emergency Preparedness and Prevention Office's Homepage at **http://www.epa.gov/ceppo/** or see Section 6.4 *Risk Management Planning* or Appendix A.3. *Summary of Principal Regulations Under the Clean Air Act* of this document. You also may obtain copies of the rule and a wide variety of technical assistance materials, as well as answers to your specific questions, from EPA's RCRA/UST, Superfund and EPCRA hotline at 1-800-424-9346 or 703-412-9810. Also check with your industry trade association for assistance in communicating with the public about risk management programs.

9.2.5   RCRA Contingency Plans

If you are a **small** or **large** quantity generator of hazardous waste, the emergency preparedness requirements under RCRA require that you develop a contingency plan for responding to spills or releases of hazardous wastes. A contingency plan will help you look ahead and prepare for accidents that could possibly occur at your food processing facility. If you are a large quantity generator (LQG), you are required to have a ***written contingency plan***. If you are a small quantity generator (SQG), you must have ***basic contingency procedures*** in place. Although a **written** contingency plan is not federally required for SQGs or conditionally exempt small quantity generators (CESQGs), it is strongly recommended. It also is important to check with your state and local authorities for any additional contingency plan or emergency preparedness requirements. See Table 8-3 in Section 8.0 *How Do I Comply With the Hazardous Waste Regulations?* for more information on contingency plan requirements for LQGs and SQGs.

> *Keep in mind that employees who are responding to releases of hazardous substances and hazardous waste are required to be trained under OSHA's Hazardous Waste Operations and Emergency Response (HAZWOPER) requirements (see 29 CFR 1910.120).*

9.3   Notification And Response Requirements

Releases or spills can occur, regardless of the amount of planning and prevention activities being conducted. The purpose of this subsection is to provide a single, general reference place of what you are required to do in response to a release or spill to the environment. Depending on the type of material released to the environment, there are various notification and response requirements under several EPA environmental statutes with which you must comply. These regulations are listed in Table 9-5.

Table 9-5. Major EPA Regulations that Address Notification
and Response Requirements

Subject	Law	Reference to Regulation
EPCRA Release Notification	EPCRA	40 CFR 355
CERLCA Release Notification	CERCLA	40 CFR 302
Notification of Spills of Oil	CWA	40 CFR 110, 116, and 117
Notification of Spills of Hazardous Substances	CERCLA	40 CFR 300 and 302
Notification of Slug Loading to POTW	CWA	40 CFR 403
Notification of discharge of hazardous waste to POTW	CWA	40 CFR 403
Notification of discharge of hazardous waste to septic system	CWA	40 CFR 144
RCRA Emergency Response	RCRA	40 CFR 262
RCRA UST Emergency Response	RCRA	40 CFR 280

Notification is the requirement to notify the appropriate authorities of a release to an environmental compartment (e.g., water, air, land). These authorities may include federal, state, and/or local government regulatory agencies, an entity with emergency planning respnsibilities such as a LEPC, or an organization responsible for responding to emergencies such as a fire department. The notification requirements are specific to each program/statute.

They are usually, but **not always**, triggered by the spill or release of a defined threshold or quantity of a substance. Knowing how these thresholds or quantities apply to your facility is critical. Sometimes, as in the release of an oil in any amount, or of a CERCLA hazardous substance in an amount that exceeds the reportable quantity, you are required to call the **National Response Center at 1-800-424-8802**. In other instances, such as the release of an EPCRA extremely hazardous substance in an amount that exceeds the reportable quantity, you are required to contact your SERC and LEPC. Table 9-6 summarizes **who** to notify and **when** for each type of notification and response requirement. (Table 9.4 provides CFR citations to each requirement.)

Response requirements specify certain procedures to be followed when responding to a spill or release, such as how to contain the release and who to contact. One of the most important aspects of the notification and response requirements is the timeframe within which the agency or emergency responder must be notified. This timeframe is typically immediately and by telephone. Several statutes require a **written follow up** notification within a specified period of time.

Table 9-6. Notification and Response Requirements

Subject	Law	Who to Notify	When
EPCRA Release Notification	EPCRA	SERC, LEPC	Immediately
CERLCA Release Notification	CERCLA	NRC	Immediately
Notification of Spills of Oil	CWA	NRC	Immediately
Notification of Spills of Hazardous Substances	CERCLA	NRC	Immediately
Notification of Slug Loading to POTW	CWA	POTW or state authorities	Immediately
Notification of discharge of hazardous waste to POTW	CWA	POTW, State Haz. Waste Authority, EPA Regional Waste Management Div. Director	Immediately; In writing
Notification of discharge of hazardous waste to septic system	CWA	EPA Regional UIC Well Program, and state UIC Program authorities	Immediately
RCRA Emergency Response	RCRA	In event of fire, call local fire department. In event of a fire, explosion or other release that could effect human health outside the facility, or if the spill has reached surface water, call the NRC.	Immediately
RCRA UST Emergency Response	RCRA	State UST permitting agency or EPA Region (whichever currently administers the UST program for your facility.	Within 24 hrs

9.3.1 EPCRA 304/CERCLA Section 103 Notification Requirements

The emergency release notification requirements set out in EPCRA and CERCLA enable federal, state, and local authorities to effectively prepare for and respond to chemical accidents. In order for a release of a EPCRA **extremely hazarous substance (EHS)** or CERCLA **hazardous substance** to be reportable, a certain amount must be released into the environment within a 24-hour period. This amount, called the **reportable quantity (RQ)**, triggers emergency release notification requirements. EHSs and their reportable quantities can be found in 40 CFR 355, Appendices A and B, and CERCLA hazardous substances and their reportable quantities can be found in 40 CFR 302, Table 302.4.

EPCRA Section 304 notification requirements are triggered for your facility if: (1) an EPCRA EHS or CERCLA hazardous chemical is **produced, used, or stored** at your facility; **AND** (2) there is a release of a CERCLA hazardous substance or EHS into the environment with a potential to affect human health and the environment offsite that equals or exceeds a reportable quantity within a 24-hour period. If a release occurs, your facility is required to **notify the SERCS and LEPCs**. See Appendix B. *Resources* for a list of SERCs and LEPCs.

Under CERCLA, if you are the person in charge of a vessel or facility and there is a release within a 24 hour period of a CERCLA hazardous substance in an amount equal to or in excess of the RQ for that substance (CERCLA 103(a), 40 CFR 302.6), you are required to immediately notify the **National Response Center (NRC) at 1-800-424-8802**. There are six specific conditions that must be met to trigger the CERCLA requirement for notifying the NRC.

Releases That Are Not Reportable. There are several types of releases that are excluded from the requirements of both EPCRA and CERCLA release notification. These releases were excluded originally under CERCLA Section 101(22) because they are covered by other regulatory programs. The regulations found at 40 CFR 355.40(a)(v) extend these statutory exclusions under CERCLA to the release reporting requirements under EPCRA.

Initial Notification. It is very important to notify the proper agency(s) and to do so as soon as practical for any reportable spill. Initial notifications can be made by telephone, radio, or in person. Under EPCRA, initial notification is required **immediately** upon discovering a spill. Thus the person making the report must use good judgement in determining how much time to spend in collecting information prior to making the notification. Specific information, such as the chemical name/identity of material(s) released, will be valuable.

Follow-up Actions for a Spill or Release. After the initial communication is established with the appropriate agencies, your facility must provide a written follow-up emergency notice, as soon as practicable after the release. The follow-up notice or notices must update information provided in the initial notice and provide information on actual response actions taken, health risks associated with the release, and advice regarding medical attention necessary for exposed individuals.

Your state also may have requirements for notifications and emergency response actions. To identify the appropriate state agencies, call the RCRA/UST, Superfund and EPCRA Hotline at 1-800-424-9346 or 703-412-9810.

9.3.2   CWA/OPA Notification Requirements

Under OPA, EPA has established requirements to report spills of oil and hazardous substances to navigable water of the U.S., including adjoining shorelines. In addition, the requirements of the CWA, OPA and RCRA, are designed to alert POTWs of impacts from spills of oil, hazardous waste or other hazardous substances to the sewer systems. They are also designed to protect septic systems from discharges of hazardous waste.

Releases of Oil and Hazardous Substances to Water

Under the Oil Pollution Prevention Regulations, you are required to meet notification requirements for releases of oil and hazardous substances to navigable water or adjoining shorelines.

Notification - The "One" Immediate Phone Call to the NRC

> **NATIONAL RESPONSE CENTER**
>
> **1-800-424-8802**
>
> **In the Washington, D.C. area:**
>
> **703-412-9810**
>
> **For more information on the NRC, access
> http://www.epa.gov/oilspill/NRC**

- Immediately notify the National Response Center (NRC) of discharges/releases of oils and hazardous substances by calling the NRC number.

- If notifying the NRC is not practicable, then immediately notify the pre-designated On-Scene Coordinator (OSC) of EPA or the USCG. (This means that you must know who your designated OSC is before the release or discharge occurs.)

- As required by the relevant Area Contingency Plan, report spills to the state, the tribal government, the territory or commonwealth where the spill occurred.

When an oil spill enters into or threatens any navigable water in the United States, coordinated teams of local, state, and national personnel are called upon to help contain the spill, clean it up, and assure that damage to human health and the environment is minimized. EPA has established requirements for reporting spills into navigable waters or adjoining shorelines. Specifically, facilities are required to report discharges of oil in quantities that may be harmful to public health or welfare or the environment.

EPA has determined that discharges of oil in quantities that may be harmful include those that: (1) violate applicable water quality standards; (2) cause a film or "sheen" upon or discoloration of the surface of the water or adjoining shorelines; or (3) cause a sludge or emulsion to be deposited beneath the surface of the water or upon adjoining shorelines.

Reporting to the National Response Center

Any person in charge of a vessel or onshore or offshore facility should notify the NRC at **1-800-424-8802** as soon as he/she had knowledge of a discharge from a vessel or facility. Spills or releases of oil which reach navigable waters or adjoining shoreline (including storm drains) or land areas which may threaten waterways must always be reported to the NRC.

When you contact the NRC, the staff person will ask you for the specific pieces of information. The NRC relays this spill information to EPA and the USCG, depending on the location of the incident. Specifically, representatives of the EPA or USCG, known as On-Scene Coordinators (OSCs), are notified. (See Section 4.6.4 Oil Spill Notification and Response for more information.)

Additional reporting. In addition, if your regulated facility experiences a single discharge of more than 1,000 gallons of oil or discharges oil in harmful quantities into or upon navigable waters in two reportable spill events during any 12-month period, you must submit a spill report in writing to the EPA Regional Administrator within 60 days.

Releases of Hazardous Substances to Water

In the case of a spill of a hazardous substance released over a 24 hour period at your facility or from facility equipment, and the released material enters a "water of the U.S." in a quantity equal to or exceeding the reportable quantity in CERCLA Section 102, you must notify the NRC as required under CWA Section 311(b); 40 CFR 116 and 117. Also note that if a spill enters a separate storm sewer that discharges to a surface water, it is subject to notification requirements. If the spilled material enters a sewer that discharges to a POTW, and it is not from a mobile source (e.g., a truck), it is not subject to these CWA notification requirements; however, you must immediately notify the POTW.

Slug Loading to POTW

Slug loading is defined as any relatively large release of a pollutant that ordinarily might not cause a problem when released in small quantities. If you know of an occurrence of slug loading at your facility that could cause problems to the POTW, you are required to notify the POTW or state immediately of a discharge of wastewater (40 CFR 403).

Hazardous Waste Sewer Discharge Notification

To make sure that hazardous wastes are not avoiding regulation by being discharged into the sewer, EPA added a provisions to the pretreatment regulations (40 CFR 403) in 1990. You must notify the POTW, your EPA Regional Waste Management Division Director, and the state hazardous waste authority (40 CFR 403) of any discharge to the POTW of a substance that would be a hazardous waste under RCRA if:

- The waste is not acutely hazardous and **more than 15 kg** (about 2.4 gallons) are discharged in a calendar month; or

•• The waste is acutely hazardous and any amount is discharged.

The hazardous waste sewer discharge notification must be in writing, and include:

- The name of the hazardous waste as listed in 40 CFR 261;
- The EPA hazardous waste number; and
- The type of discharge (e.g., "batch" for a single event spill, such as a drum or container; or "continuous" for a large spill that has not stopped).

If **more than 220 lb** (100 kg, or approximately 25 gallons) of hazardous waste is discharged to the sewer per month by your food processing facility, then you also must include the following information in the notification:

- The hazardous constituents in the waste;
- An estimate of how much hazardous waste (mass and concentration) was discharged to the sewer during that month; and
- An estimate of how much hazardous waste you will discharge in the next 12 months.

You should keep the telephone numbers of the people that you must notify (e.g., the POTW, your EPA Regional Waste Management Division Director, and the state hazardous waste authority) at the facility. Call your EPA and state regulatory agencies to get the appropriate contact numbers.

Hazardous Waste Septic System Notification

If the discharge of any amount of a hazardous waste is to a septic system, you must immediately notify the EPA Regional Underground Injection Control Program and the state Underground Injection Control Program. Call your EPA and state regulatory agencies to get the appropriate contact numbers.

9.3.3 RCRA Emergency Response Requirements

Your RCRA contingency plan should tell you what to do if you have an accidental or emergency release of a hazardous waste at your food processing facility, and what to do in case of emergencies such as fires or explosions (see Section 8.0 *How Do I Comply With the Hazardous Waste Regulations?* and Section 9.2.5 *RCRA Contingency Plans* for more information). In the event of a hazardous waste release, RCRA emergency response requirements

Under RCRA, materials used in cleanup operations following a hazardous material or oil spill are considered hazardous wastes. These cleanup materials are considered part of your total monthly accumulation and may affect your generator status. See Section 8.0 for information on determining generator status.

contain the following procedures for responding to a spill or release of hazardous waste(s):

- Contain the flow of hazardous waste to the extent possible, and as soon as is possible, clean up the hazardous waste and any contaminated materials or soil.

- In the event of a fire, call the fire department and, if safe, attempt to extinguish the fire using a fire extinguisher. After the fire is out, contain the release as described above.

- In the event of a fire, explosion, or other release that could threaten human health outside the facility, or if you know that the spill has reached surface water, follow the instructions provided in Section 9.3.2.

All employees must know proper waste handling and emergency procedures. In addition, a person at your facility must be appointed to act as the emergency coordinator to ensure that emergency procedures are carried out properly. The responsibilities of the emergency coordinator are: (1) he/she will be available 24 hours a day (at the facility or by phone); and (2) he/she will know whom to call and what steps to follow in an emergency. See Section 8.0 *How Do I Comply With the Hazardous Waste Regulations?* for more information.

> *Keep in mind that employees who are responding to releases of hazardous substances and hazardous waste are required to be trained under OSHA's Hazardous Waste Operations and Emergency Response (HAZWOPER) requirements (see 29 CFR 1910.120).*

9.3.4 RCRA UST Emergency Response Requirements

RCRA includes emergency response requirements for leaking underground storage tanks (USTs) (in 40 CFR 280.53), including reporting, response, and cleanup procedures. If your facility has USTs that contain petroleum or hazardous substances, and you identify any of the following conditions associated with your UST(s), you must report:

- Unusual operating conditions exist (e.g., erratic behavior of product dispensing equipment, sudden loss of product from the UST system, or an unexplained presence of water in the tank) unless due to defective but not leaking equipment;

- Monitoring results (see Section 8.0) indicate that a release has occurred; or

- Regulated substances are observed or discovered at the UST site (e.g. free vapors in the soils, basements, sewer and utility lines, and/or a sheen on nearby surface waters).

Your report should be made within 24 hours to the state UST permitting agency or the EPA Region, whichever currently administers the UST program for your facility. To help identify who to contact, call EPA's RCRA/UST, Superfund and EPCRA Hotline at 1-800-424-9346 or 703-412-9810 or visit EPA's Office of Underground Storage Tanks website at **http://www.epa.gov/ OUST/**.

In addition to this report, RCRA (40 CFR 280) requires that you immediately contain and clean up a release from an UST that contains:

- Petroleum, where the spill exceeds 25 gallons or causes a sheen on a nearby surface water, or is less than 25 gallons but cannot be cleaned up within 24 hours.

- A CERCLA hazardous substance (listed at 40 CFR 302.4, Table 302.4) above the reportable quantity, or below the reportable quantity but cannot be cleaned up within 24 hours.

Following notification, response actions required for leaking USTs include taking immediate action to prevent any further release of the regulated substance into the environment, and identifying and mitigating fire, explosion, and vapor hazards. The owner/operator must submit a report summarizing initial abatement measures (usually within 20 days) including:

- Removal of the regulated substance from the UST;

- Inspection of aboveground or exposed below ground releases and preventing migration of the substance into surrounding soils and ground water;

- Continued monitoring and mitigating safety hazards;

- Remedying hazards posed by contaminated soils that have been excavated or exposed; measuring for the presence of a release where contamination is most likely to exist.

Several follow-up procedures (initial site characterization, free product removal, and investigations for soil and groundwater cleanup, and corrective action plan) may be required (see 40 CFR 280.63 - 280.66). State requirements for response and clean up activities vary; therefore be sure you contact the appropriate implementing agency (state or EPA region) for your facility for additional requirements that may apply. See Section 8.0 *How Do I Comply With the Hazardous Waste Regulations?* for more information.

9.4 Summary

This section has highlighted some of the similarities, differences, and complexities of EPA's planning, reporting, notification, and response requirements under the various federal statutes. The complexity stems, in part, from the differences in statutory approach. Some statutes have requirements that pertain to spills or releases to specific environmental compartments (e.g., water, air, land), whereas other statutes have requirements that focus on specific substances, irrespective of the environmental compartment to which they are released.

It is very important to understand the differences as well as the interconnections among these requirements in order to prepare the appropriate plans for your facility, complete the required reporting activities for hazardous materials that you have onsite, and respond appropriately to releases and spills. In addition, you may reduce your liability by preventing potential releases or responding properly in the event of a release. Don't rely solely on the information in this section to meet these requirements. You must review the regulations thoroughly to figure out your responsibilities and comply with them. Contact your EPA and state regulatory agencies for assistance and additional information.

SECTION 10 CONTENTS

10. OTHER MAJOR ENVIRONMENTAL STATUTES AND REGULATIONS: CERCLA, RCRA SUBTITLE D, FIFRA AND TSCA

In addition to the major statutes discussed previously in this guide, there are other environmental statutes and regulations that you must comply with as a food processing facility. These include:

- Comprehensive Environmental Response, Compensation, and Liability Act (CERCLA) requirements apply to all food processing facilities that release hazardous substances into the environment.

- Subtitle D of the Resource Conservation and Recovery Act (RCRA). Subtitle D requirements apply to all food processing facilities that dispose of solid, nonhazardous wastes.

- Federal Insecticide, Fungicide, and Rodenticide Act (FIFRA). FIFRA requirements apply only to those facilities that apply and store pesticides, such as herbicides, insecticides, and rodenticides.

- Toxic Substances and Control Act (TSCA). TSCA requirements apply to facilities subject to the TSCA Chemical Inventory Update, and may apply to those facilities that manage substances such as asbestos, chlorofluorocarbons (CFCs), and polychlorinated biphenyls (PCBs).

10.1 Comprehensive Environmental Response, Compensation, and Liability Act

Under CERCLA, commonly known as Superfund, EPA can respond to releases, or threatened releases, of hazardous substances that may endanger public health, welfare, or the environment. Under the **hazardous substance release reporting regulations** of CERCLA Section 103, (40 CFR 302), your facility is required to report to the National Response Center (NRC) any release into the environment of a hazardous substance that exceeds the reportable quantity for that substance. More than 700 hazardous substances are subject to the emergency notification requirements under CERCLA Section 103(a) (40 CFR 302.4) as well as those on the list of "extremely hazardous substances" under EPCRA (40 CFR 355). See Section 7.0 for more information about both EPCRA and CERCLA notification requirements.

In response to releases, EPA implements **hazardous substance responses** according to procedures outlined in the National Oil and Hazardous Substances Pollution Contingency Plan

(NCP) (40 CFR 300). The NCP includes provisions for permanent cleanups, known as remedial actions, and other cleanups referred to as "removals." While EPA generally takes remedial actions only at sites on the National Priorities List (NPL), which currently includes approximately 1300 sites, both EPA and states can act at other sites. In addition, EPA can force parties responsible for environmental contamination to clean up the contamination or reimburse the Superfund for response or remediation costs incurred by EPA. EPA encourages community involvement throughout the Superfund response process.

Note: EPA's RCRA/UST, Superfund and EPCRA Hotline at 1-800-424-9346 or 703-412-9810 provides information and references guidance pertaining to the Superfund program.

10.2   Subtitle D of the Resource Conservation and Recovery Act

Subtitle D of RCRA and its implementing regulations basically apply to the management of solid, nonhazardous waste and its disposal in landfills. See Section 8.0 *How Do I Comply With the Hazardous Waste Regulations?* for a definition of solid waste. A nonhazardous waste is defined as any garbage, refuse, or sludge from waste treatment plants, water treatment plants, or air pollution control equipment. Some examples of nonhazardous food processing solid wastes include discarded cardboard, food wastes, waste papers, and other food packaging materials. A description of typical solid wastes generated by food processing operations is presented in Section 3.0, Table 3-2 *Waste Analysis for SIC Code 203 Facility* of this guide. While this list is not all inclusive, it gives you a general idea of the kinds of solid wastes generated for each operation.

Subtitle D applies to your food processing facility because it prohibits open dumping of solid, nonhazardous wastes. Programs addressing the disposal of solid, nonhazardous wastes are developed and enforced at the state or local level. You should contact your state regulatory agency or a local reputable waste contractor for more information on proper disposal practices.

To reduce the volume of solid, nonhazardous wastes requiring land disposal, you are encouraged to recycle or reuse as many waste materials as possible. Keep in mind that all pollution prevention activities should be carried out in accordance with food safety requirements. Many states have recycling programs in place for a variety of waste materials, particularly glass, plastic, paper, and cardboard. You should contact your State Pollution Prevention Office to get information on the recycling programs available in your area. In addition to state programs, there may be local recycling requirements. Check with your local regulatory agency for more information. See Appendix B *Resources* for information on state contacts, etc.

10.3 Federal Insecticide, Fungicide, and Rodenticide Act

This section describes the requirements for managing pesticides under the Federal Insecticide, Fungicide, and Rodenticide Act (FIFRA), as well as the requirements under the Food Quality Protection Act (FQPA).

10.3.1 Use of Pesticides in the Food Processing Industry

FIFRA primarily regulates the manufacture and registration of pesticides (40 CFR Parts 152 and 156), but important requirements also exist for pesticide **users**. Your food processing facility may at some time store, apply (or have applied), and dispose of pesticides. There are many types of pesticides, including herbicides, insecticides, rodenticides, and antimicrobial pesticides (e.g., disinfectants, sanitizers). Pesticides must be applied only according to label directions established by EPA. Using a pesticide in a manner inconsistent with its labeling constitutes misuse and is illegal [FIFRA Section 2 (ee)].

Pesticides can be used to control a variety of pests that are associated with food processing facilities in the U.S., including:

- Birds (e.g., English sparrows, pigeons, and starlings)
- Weeds
- Rodents (e.g. house mouse, rat, and roof rat)
- Insects (e.g., cockroaches, moths, beetles, flies, ants).

These pests can be controlled using direct application of the appropriate avicide, herbicide, rodenticide, or insecticide; or by fumigants. Fumigants are chemicals that are in the gas phase at effective temperatures, and they penetrate cracks, crevices, and the commodity being treated. Fumigants, while toxic to insects, rats, birds, mammals, weed seeds, nematodes and fungi, are also highly toxic to humans and may leave toxic

> *Herbicides can be used to eliminate or inhibit tree and weed growth around facilities, while insecticides and rodenticides may be used to control pests.*

residues or tastes or odors. Fumigants can be applied by several methods, are readily available, and are economical to use. They must be applied with the proper protective equipment and by certified applicators.

Antimicrobial pesticides comprise a broad range of products designed to control undesirable microorganisms such as bacteria, viruses, or algae on non-living objects (inanimate) or surfaces, and on raw fruits and vegetables (FIFRA Section 2(mm)(1)(A)). Some antimicrobial pesticides are used to sterilize, disinfect, or sanitize certain items, including food preparation areas. While primarily regulated under FIFRA, the FQPA changes the jurisdiction of some antimicrobial products from FIFRA to the FQPA (see Section 10.3.2). Since late 1996, the

Antimicrobials Division within EPA's Office of Pesticide Programs (OPP) has been responsible for all activities related to the regulation of antimicrobial pesticides. For more information on antimicrobial pesticides, access the OPP website at **http://www.epa.gov/oppfead1/cb/ csb_page/qsas/antimic.htm/** or contact the Antimicrobial Division Ombudsman at 1-800-447- 6349.

Requirements for Pesticide Use

FIFRA requires that all pesticides be registered for every intended use, and that labels containing instructions for proper storage, use, and disposal accompany each pesticide marketed. Application and handling requirements are specific to each pesticide product. Excess pesticides that must be disposed of may, in some cases, be considered hazardous waste, and must be managed accordingly (see Section 8.0 *How Do I Comply With the Hazardous Waste Regulations?*). Under FIFRA, the use of pesticides in a manner inconsistent with labeling established by EPA is illegal. You can be held responsible if any pesticides applied on your property are misapplied or mishandled. The "label is the law."

RUPs: When Your Facility Contracts for Pesticide Application. While some facilities may elect to hire a contractor for all of their pesticide applications, all facilities may have to contract out pesticide applications at one time or another. Under FIFRA, some pesticides, which are referred to as restricted use pesticides (RUPs), have been deemed by EPA to have high toxicity or to pose particular environmental hazards. Pesticide labels will clearly state whether a particular pesticide is restricted use only.

> *When a pesticide is applied by a contractor, the contractor **and** the person contracting for the service may be held responsible for pesticide misuse.*

RUPs may only be applied by **certified** pesticide applicators; there are two types: "commercial" and "private." A **commercial** applicator is certified to apply pesticides to other people's property. Unless your facility chooses to certify some employees in pesticide application, applications of RUPs will require the use of a certified commercial applicator.

You should always verify that your contractor, or in the case of RUPs, the certified commercial applicator, uses the correct pesticide application rate and method. The pesticide label contains detailed information on appropriate rates and methods of application. The actual application should be observed to ensure that application methods are correct.

Non-RUPs: When Your Facility Applies Pesticides. For pesticides that are not restricted use (non-RUPs), you may purchase, store, apply, and dispose of these pesticides. Your selection of pesticide(s) should be based on the type of pests or weeds to be controlled, and the most environmentally sound applications. Best management practices (BMPs) for pesticide application include selecting pesticides with low mobility or toxicity to protect both humans and the environment, and use of pesticides that target

> ***Selecting Pesticides***. *Your local agricultural extension office can provide guidance when selecting the most appropriate pesticide to use. In addition, pesticide labels provide detailed information as to the appropriate use of a pesticide.*

individual pests or weeds. Alternatives should be considered when selecting a pesticide such as those that require the minimum amount of active ingredient to be applied to control a problem. See *Federal Register* notice 58 FR 26856.

Pesticide Storage. EPA has published guidelines for safe storage of pesticides and EPA may place storage conditions on a pesticide's label.

Application or Use of Pesticides. FIFRA requires that every pesticide be registered and labeled with both the appropriate application methods and the appropriate amounts to be used in a particular application. Pesticide application includes mixing and application of the pesticide. It is a violation of FIFRA to apply a pesticide in a manner inconsistent with its label. Therefore, carefully read the label of any pesticide used and use only the amounts specified by the label. To minimize potential environmental impacts, the minimum application rate that is effective should always be used. Section 2(ee) of FIFRA does allow for some variances to the label requirements.

> *Inventory: You should plan and order only the amounts of pesticides needed at the time of application. Manufactures may allow you to return unused or unopened products.*

Mixing should be conducted at a mixing site where structures exist to contain any spills. Check with your state before constructing pesticide mixing or loading sites to learn about state requirements affecting the location or placement of such a site.

Post-Application Clean Up and Pesticide Disposal. After pesticides are used, application equipment must be cleaned and empty containers disposed of. Consistent with the EPA label, some pesticide containers may be disposed of in municipal solid waste landfills; other containers may be disposed of in a licensed landfill or incinerator. The rinse water from clean up, if not reused, may be considered hazardous and should be disposed of accordingly. Disposal of **unused** pesticide product depends on the type of pesticide and EPA requirements on the label, such as incineration in a pesticide incinerator or other special treatment, or encapsulation and disposal at a properly licensed facility. EPA has proposed special requirements for the disposal of recalled or canceled pesticides under FIFRA Section 19, *Pesticide Management and Disposal*, 58 FR 26856, May 5, 1993. EPA expects to finalize these requirements in the Fall 1998.

> *Be sure to ask your disposal facility if they are licensed to accept the type of pesticide wastes you are disposing.*

Pesticide Use/Applicator Training. As noted above, certain pesticides are classified by the EPA as restricted use (RUPs) based on toxicity or environmental hazard. These pesticides may be applied only by a licensed certified applicator. EPA sponsors a Pesticide Applicator Training Program that is administered by the states, primarily through local agricultural extension offices. Contact your local extension office to receive training in pesticide application to become a licensed certified applicator.

Pesticide **worker protection standards** (WPSs) promulgated by EPA require that pesticide workers receive training in the proper application of pesticides within five days of entering an area where pesticides are being applied. EPA does not require right-of-way workers to comply with the WPS. However, it is good practice for employees working with pesticides to receive training to ensure that pesticides are applied properly.

Recordkeeping. Best management practices (BMPs) for pesticides include keeping accurate records of inventory, use, and storage for the following purposes: tracking when the next application should occur in accordance with label directions; managing inventory; responding in the event of an accidental spill or fire; and alerting emergency responders of stored pesticides.

10.3.2 Food Quality Protection Act

The Food Quality Protection Act (FQPA) of 1996 was a comprehensive overhaul of the laws that regulate pesticides in food: FIFRA and the Federal Food, Drug and Cosmetics Act (FFDCA). The new law amends both major pesticide laws to establish a more consistent, protective regulatory scheme.

EPA's Role. EPA plays a role under both of these statutes in regulating pesticides:

- Under FIFRA, EPA registers pesticides for use in the United States and prescribes labeling and other regulatory requirements to prevent unreasonable adverse effects on human health or the environment.

- Under the FFDCA, EPA establishes tolerances (maximum legally permissible levels) for pesticide residues in food. Tolerances are enforced by the U.S. Department of Health and Human Services (USDHHS)/FDA for most foods, and by the USDA/FSIS for meat, poultry, and some egg products.

Changes from FQPA. The FQPA made many revisions to the way pesticides are regulated:

Under the FFDCA, the new law establishes a health-based safety standard for pesticide residues in food; requires an explicit determination that tolerances are safe for children; sets limitations on benefits considerations; requires tolerance reevaluation; incorporates provisions for endocrine testing; includes enhanced enforcement of pesticide residue standards by allowing the FDA to impose civil penalties for tolerance violations; requires distribution of a brochure in grocery stores on the health effects of pesticides, how to avoid risks, and which foods have tolerances for pesticide residues based on benefits considerations; and does not allow states to set tolerance levels that differ from national levels unless the state petitions EPA for an exception, based on state-specific situations.

Under FIFRA, the new law requires tolerances to be reassessed as part of the reregistration program; requires EPA to periodically review pesticide registrations to ensure that all pesticides meet updated safety standards; expedites review of safer pesticides to help them reach the market sooner and replace older and potentially more risky chemicals; establishes minor use programs within EPA and USDA to foster coordination on minor use regulations and policy; and establishes new requirements to expedite the review and registration of antimicrobial pesticides. While some of these changes under the FQPA do not affect you directly as a regulated entity, it is to your benefit to be aware of these changes in the regulations affecting pesticides residues.

> Note: The FQPA changed the jurisdiction of some antimicrobial products from FIFRA to the FQPA.

Environmental Stewardship. You should also be aware that there are opportunities for environmental stewardship. EPA's Pesticide Environmental Stewardship Program (PESP) is a voluntary program dedicated to protecting human health and the environment by reducing both the use of pesticides and the risks associated with pesticide use. Current partners include agricultural producers as well as non-agricultural interests. Partners in PESP volunteer to develop and implement a well-designed pesticide management plan that will result in the safest and most effective way to use pesticides. In turn, EPA provides a liaison to assist the partner in developing comprehensive, achievable goals. Liaisons act as customer service representatives for EPA, providing the partner with access to information and personnel. EPA also will attempt to integrate the partners' stewardship plans into its agricultural policies and programs. For more information, call the PESP Hotline at 1-800-972-7717 or access the PESP webpage at **http://es.epa.gov.partners/pest/pest.html/**.

10.4 Toxic Substances Control Act

Under TSCA, EPA collects data on chemicals in order to evaluate, assess, mitigate, and control risks which may be posed by their manufacture, processing, and use. TSCA provides a variety of control methods to prevent chemicals from posing unreasonable risk, and the standards may apply at any point during a chemical's life cycle. Some food processing may be subject to the TSCA Chemical Inventory Update (see below) based on the type and quantity of substances they manufacture. Facilities may also be subject to requirements for asbestos, CFCs, and PCBs.

Regulated Substances under TSCA. You should be aware that drugs, cosmetics, foods, food additives, pesticides, and nuclear materials are **exempt from TSCA** and are subject to control under other federal statutes (e.g., foods and food additives are under the purview of the Federal Food, Drug and Cosmetics Act (FFDCA) administered by the FDA. In order for a food or food additive to be exempt, however, it must meet the definition contained in the FFDCA (21 USC 321 et seq.), or related statutes such as the Poultry Products Inspection Act and the Federal Meat Inspection Act. If the food or food additive does not meet the definition, the substance may then be regulated under TSCA and is subject to all the requirements of TSCA including testing, premanufacture notice, reporting and recordkeeping, export notification, and import certification. For example, vegetable oils and their derivatives from vegetable processing that are used as an ingredient in lubricants, paints, inks, fuels, plastics, solvents and a variety of other industrial products are subject to all of TSCA's requirements.

TSCA Chemical Inventory Update Reporting. Under TSCA Section (8)(a), manufacturers and importers of certain chemical substances are required to report the specific chemical identity and quantity, and site of manufacture or importation of these substances every four years. This is known as the *TSCA Chemical Inventory Update*; and the next four-year reporting period begins in August 1998 and ends in December 1998.

If you are manufacturing substances, such as vegetable oil and animal fats, that are used for non-food purposes (e.g., in inks), you must comply with requirements of the Inventory Update. Certain exemptions are available to small manufacturers under 40 CFR 710.28. EPA uses this information to update its *TSCA Chemical Substances Inventory* database. EPA relies on the accuracy of this data to monitor and estimate health and safety risks to people and the environment as well as to formulate control and preventive responses. Note: Other EPA programs may develop site-specific information on public and environmental risks as needed. EPA also incorporates reported information into its regulatory decision-making process to assure responsive and effective regulation of the chemical industry.

To determine your reporting obligations, you must make two determinations for each substance that you manufacture in the United States or import into the United States:

(1) Is the substance reportable under the Inventory Update Rule? If you do not already know whether a substance you manufacture or import is on the Inventory, you should consult a copy of the latest version of the inventory. To obtain a copy (at a cost), call the Superintendent of Document, Government Printing Office at (202) 783-3238, or call the TSCA Assistance Information Service at 202-554-1404.

(2) Are you a manufacturer, importer, or exporter who is required to report that substance? Generally, if you manufacture or import 10,000 lbs or more of a reportable substance at any single site during the fiscal year preceding the reporting period, you are required to report. Small manufacturers are usually exempt from the reporting under the Inventory Update Rule; however, there are some conditions under which they must report.

There are requirements for when and how to report the information for the Inventory Update Rule. There are also recordkeeping requirements. You must maintain records that document the information contained in your submissions. Required records include those that show the production volume, plant site, and site-limited status of each substance reported. These records must be kept for four years after the effective date of the applicable reporting period.

If you are manufacturing or importing a chemical substance that is not already on the inventory (and has not been excluded by TSCA), you must submit a premanufacture notice (PMN) prior to manufacture or importation (40 CFR 720). The PMN must identify the chemical and provide available information on health and environmental effects. If available data are not sufficient to evaluate the chemical's effects, EPA can impose restrictions pending the development of information on its health and environmental effects. EPA also can restrict significant new uses of chemicals based upon factors such as the projected volume and use of the chemical.

Reporting and Recordkeeping Requirements. Section 8 of TSCA authorizes EPA to require chemical manufacturers, importers, and processors to keep records and report certain information. TSCA Section 12 requires the submission to EPA of certain information about chemical exports, while Section 13 requires the submission of certification statements concerning import shipments of chemical substances. These additional requirements under TSCA are summarized below:

- **Alleged Significant Adverse Reactions.** Under TSCA Section 8(c), if you manufacture, import, process, or distribute chemical substances or mixtures in commerce, you are required to keep files of allegations of significant adverse reactions and provide this information to EPA upon request.

- **Health and Safety Studies Submission.** Under TSCA Section 8(d), if you manufacture, import, process, or propose to manufacture, import, process listed chemicals, you are required to submit lists or copies of unpublished studies to EPA.

- **Substantial Risk Reporting.** Under TSCA Section 8(e), if you manufacture, import, process, or distribute a chemical substance or mixture and obtain "new" information which reasonably supports the conclusion that such substance or mixture presents a substantial risk of injury to health or the environment, you are required to report such information to EPA within 15 days.

- **Exports.** Under TSCA Section 12, if you export chemicals subject to final and certain proposed rules and orders under TSCA Sections 4, 5, 6 and 7, you are required to notify EPA of the country of destination the first time a chemical is shipped to that country during a calendar year.

- *Imports.* Under TSCA Section 13, if you import chemical substances, you are required to certify that each shipment is in compliance with TSCA or is not subject to TSCA.

Asbestos, CFCs, and PCBs. There may be other substances, including asbestos, CFCs, and PCBs, at your food processing facility which are regulated under Section 6 of TSCA. Because these substances pose unreasonable risks, EPA can ban the manufacture or distribution in commerce, limit the use, require labeling, or place other restrictions on these substances (40 CFR 750). If you have these substances onsite, you should check with your EPA and/or state regulatory agencies for information on federal and/or state requirements for these substances. For additional assistance, contact the TSCA Assistance Information Service at 202-554-1404.

SECTION 11 CONTENTS

11. POLLUTION PREVENTION TECHNIQUES

11.1 Introduction

Pollution prevention (P2) is a simple idea: it means you eliminate pollution *before it is created* at your food processing facility rather than controlling the pollution from your processes and then treating and disposing of the wastes that you generate. P2 techniques that food processing facilities can use range from placing catch pans near equipment hydraulic lifts to making fundamental changes in the way food is cleaned and prepared. This section discusses the benefits and incentives, costs of compliance, and techniques that may work at your facility. *Keep in mind that all P2 activities should be carried out in accordance with food safety requirements of the U.S. Department of Agriculture (USDA) and the Food and Drug Administration (FDA).*

The U.S. Environmental Protection Agency (EPA) defines P2 as the use of materials, processes, or practices that reduce or eliminate the generation of pollutants or waste at the source. The direct benefits of P2 are:

- Decreased waste management costs
- Decreased input materials costs and energy consumption
- Decreased environmental compliance costs
- Decreased liability
- Increased compliance
- Increased worker safety
- Improved corporate image.

What will these benefits mean to your food processing facility?

- *Reduction in the cost of operating your food processing facility*

    The creation of waste that impacts the water, land, or air, and the use of certain chemicals, translates into additional dollars you must spend. When you generate waste, your operating costs increase since you must pay for items, such as hazardous waste disposal, the installation and operation of pollution control equipment, and permit fees. By reducing wastestreams, you can cut the cost of operating your facility. And these cost savings should translate to **lower operating costs and increased profits**.

- *A more efficient and productive business*

    In order to maintain compliance with environmental regulations, you and your staff must conduct a great number of environmental management activities. These activities cost your facility time and money. More often than not, these costs are hidden in your facility's overhead. The more waste you generate, the more your facility is regulated. So, if you spend less time on compliance activities because you have less waste to manage, your facility will have more time to process foods.

*   ***Reduced Risk of Liability***

    You will decrease your risk of liability by reducing the volume and the potential toxicity of the vapor, liquid, and solid discharges you generate. As a food processing facility, you should look at all types of waste, not just those that are currently defined as hazardous (see Section 8.0 *How Do I Comply With the Hazardous Waste Regulations?* for a definition of hazardous waste). Since toxicity definitions and regulations change, reducing volumes of wastes in all categories is a sound long-term management policy.

*   ***Prevent pollution***

    If there are fewer hazardous materials at your food processing facility, your compliance obligations will be fewer. If your workers are exposed less frequently to hazardous materials, their health and safety will not be as much at risk. In addition, you will not have to be concerned about their well being -- or your liability. Furthermore, the environment will be cleaner and you will be prepared for a regulatory agency's inspection.

> *Successful implementation of pollution prevention techniques can reduce worker exposure and liability.*

11.2 What Pollution Prevention Techniques Can I Use?

This section presents an overview of P2 techniques that can be incorporated into your major process activities (e.g., storage, receiving and preparation, processing and filling, packaging, and storage and distribution), as well as your ancillary operations (e.g., refrigeration, cleaning, maintenance, and laboratory activities). The techniques shown in *Table 11-1 Overview of Pollution Prevention Techniques* provide a general overview of several of the options available to you.

Section 11.3 Pollution Prevention Techniques for the Food Processing Industry presents detailed descriptions of each P2 technique. It is important to remember that not every P2 technique will work at every food processing facility. You should compare and evaluate these P2 techniques to identify those that may help you meet your P2 goals. You will then need to try a select few to determine what works in your facility, but does not compromise the quality and safety of your product. *Consultation with the agencies regulating food safety is critical during the planning and evaluation of any pollution prevention technique(s) that you may adopt.*

> *Some P2 techniques will assist you in reducing your fresh water use and wastewater generation. This will result in cost savings to your facility and decreased demands on the POTW to process your wastewater.*

As shown in Table 11-1, there are many different kinds of P2 techniques. These techniques can be divided into categories, including process or equipment modification (primarily involving utilizing water conservation methods); operational and housekeeping changes; recycling/reuse; and material substitution and elimination. For the purposes of this document, each technique is placed under one of these categories. However, you may categorize a particular technique

Table 11-1. Overview of Pollution Prevention Techniques

Type of P2 Technique	Technique	Process or Ancillary Activity	Ease of Implementation
Process/ equipment modification	Replacing traditional faucets	Receiving and preparation	Easy - Moderate
	Dry caustic peeling of fruits and vegetables	Receiving and preparation	Difficult
	Water shutoff during breaks	Processing and filling	Easy
	Water control units	Processing and filling	Moderate
	Installing flow meters	Processing and filling	Easy
	Exterior area water use reduction	Storage and distribution	Easy
Operational and housekeeping changes	Placing catch pans under potential overflows/leaks	Storage	Easy
	Covering outside storage areas	Storage	Easy
	Inspections and preventive maintenance of potential discharge areas	Storage	Easy
	Secondary containment	Storage	Easy - Moderate
	Monitor liquid fill machines	Processing and filling	Easy - Moderate
	Covering outside drains during loading and unloading	Storage and distribution	Easy
	Covering inside floor drains (in non-production areas only)	Maintenance	Easy
	Cleaning prevention	Cleaning	Easy - Difficult
	Precleaning and dry cleanup	Cleaning	Moderate
	Skim grease traps regularly	Cleaning	Easy
	Screening	Cleaning	Moderate
	Minimizing pests	Cleaning	Easy - Moderate
Recycling/reuse	Countercurrent washes	Processing and filling	Moderate
	Process water reuse	Processing and filling	Easy - Moderate
	Water recirculaton units	Processing and filling	Moderate
	Water used to chill products	Processing and filling	Moderate
	Residuals management	Processing and filling, storage and distribution	Easy - Moderate
	Recycling refrigerants	Refrigeration	Moderate
	Reducing/recycling/reusing packaging	Processing and filling	Easy - Moderate

Table 11-1. Overview of Pollution Prevention Techniques

Type of P2 Technique	Technique	Process or Ancillary Activity	Ease of Implementation
Material substitution and elimination	Laboratory inventory reduction	Laboratory	Easy
	General inventory control	Purchasing	Easy
	Using alternative refrigerants	Refrigeration	Moderate

differently for your operation. The table also indicates the ease of implementation of each technique. While some P2 techniques are easy; others are more challenging. However, they all involve changes in how you do business. When you understand how much it costs to comply with all the regulations that apply to your facility, you will see that changing your operations makes good business sense.

11.3 Pollution Prevention Techniques for the Food Processing Industry

This section describes P2 opportunities that could be implemented at your facility. Information on whether the technique is easy or more difficult to use is included next to each listing, followed by a description of the technique. The ease of implementation can be determined by many factors, such as cost, adding new equipment, substituting materials, and if necessary, making associated process changes. *Food processors should evaluate these P2 techniques before use to assure they do not compromise the safety of their product.*

11.3.1 Techniques for Process/Equipment Modification

Replacing Traditional Faucets   Easy- Moderate

As a food processing facility, you have probably found that traditional faucets can be one of the highest water users in your facility. Traditional faucets are often large water users because they have a high flow rate, and they can be left on while unattended, sometimes for hours at a time. By replacing the faucets with modified flow faucets, flow rates can be reduced by over 80%. By retrofitting faucets with on-demand foot or knee control devices or automatic shutoff nozzles, flow can be reduced even further. An example of such savings is presented below.

At a Kentucky Poultry plant[1], 44 faucets were replaced and upgraded leading to an annual savings of $37,174. The plant's cost of installing 44 restricted flow faucets was $1,100 at $25 per faucet. The new faucets had flow rates of 0.5 gpm compared to 1.5 - 3.5 gpm for the old faucets. The change reduced the process line's flow rate by 83.5 gpm (from 87.5 gpm to 4 gpm). Total savings were calculated as follows:

*83.5 gpm x 60 min/hr x 16 hr/day (work day) x 265 days/yr (operating days) = 21,424,400 gal/yr x $1.75/1,000 gal = **$37,174/yr savings***

$37,174/yr / 265 days/yr = $101.85/day
*$1,100(total cost) 101.85/day = **11 day payback period**.*

Note: An additional step for water conservation can be the use of automatic shutoff valves which can stop sprays when conveyor belts stop.

Dry Caustic Peeling of Fruits and Vegetables   Difficult

As a food processing facility, you may have problems with high levels of product residue in the water generated during the steam peeling process. In conventional steam peeling operations, potato peels may contribute up to 80 percent of the total plant wastewater biochemical oxygen demand (BOD). However, peeling processes can be modified so that the peel waste can be removed without using excessive amounts of water. One option is the "dry" caustic peeling process.

In a dry caustic system, peels are softened by caustic, and then a machine uses very thin soft rubber discs to remove the peels. These rubber disks are placed on rotating cylindrical rolls arranged in a circular revolving cage containing a feed screen through the center. The feed rate is controlled by the central screw conveyor. A final rinse to remove the last traces of peel and caustic is the only fresh water used.

Table 11-2 Comparison of the Average Liquid Effluent for Caustic and Dry Peeling Operations presents a comparison of effluent from conventional caustic and dry caustic peeling operations, based on a demonstration of peach peeling at a canning facility.

---

[1] U.S. Environmental Protection Agency. Climate Wise - Economic and Environmental Impact Case Studies: Food Processing. "Case Study: Waste Reduction Opportunity Assessment. Seaboard Kentucky Poultry Processing Facility. Hickory, Kentucky. February 1994."

Table 11-2. Comparison of the Average Liquid Effluent for Caustic and Dry Peeling Operations (Del Monte Demonstration Project)[1]

Wastewater Characteristics	Conventional Caustic Peeling	Dry Caustic Peeling
Water Usage	850 gallons/ton[2]	90 gallons/ton
COD	10.8 (1500 ppm)	4.2 (5600 ppm)
BOD	6.7 (940 ppm)	2.8 (3700 ppm)
Suspended Solids	5.6 (790 ppm)	1.9 (2500 ppm)
Total Solids	17.8 (2500 ppm)	4.0 (5300 ppm)
pH range	6-9	4-6

[1] Carawan, Roy et al., "Spinoff On Fruit and Vegetable Water and Wastewater Management," presented in *Industrial Water Conservation References of Food Processing*, California Department of Water Resources, 1989.

[2] Assumes countercurrent rinse. Without countercurrent rinse, this number could be as high as 2,000 gallons per ton for peaches.

Water Shutoff During Breaks   Easy

If your food processing facility does not have on demand faucets and hoses, water shut off during breaks can save thousands of dollars each year, without any capital investment. For example, shutting off water during breaks at the Kentucky poultry plant discussed earlier saved $23,964 per year. Based on its previous water use of 344.5 gpm during breaks, its savings were calculated as:

344.5 gpm x 60 min/hr x 2.5 hr/day (break time) x 265 day/yr = **13, 693,875 gal/yr**
13, 693,875 gal/yr x $1.75/1,000 gal = **$23,964/yr**.

Water Control Units   Moderate

Your food processing facility may provide a continuous flow of fresh water for the raw product prior to and during preparation, or you may require continuous replenishment of a wash bath for each new batch of product. A water control unit can be added to the automatic process to reduce fresh water use. Wall-mounted control units, which control the flow and temperature of the water to the wash bath, can be installed. A water control unit costs approximately $1,200. The benefits of this technique are in the cost savings which can be realized from decreased fresh water use and reduced wastewater discharge.

Installing flow meters   Easy

When combined with education and training, flow meters can help all employees become involved in your facility's water reduction program. Food processing facilities have found that flow meters allow them to measure and monitor water use on a constant basis. This technique is especially useful in cooking operations, where any excess water that enters the process is excess water that is heated. Thus by preventing excess water from entering the process, you can save energy costs of heating excess water. Flow meters allow all employees to monitor water use and help reduce water usage on a facility-wide basis.

Exterior Area Water Use Reduction   Easy

In addition to the pollution prevention techniques directly related to your production process, you have additional opportunities to reduce water usage. By educating all employees about the costs of water use and the benefits of reduction, your facilities can maximize cost savings. Some options for reducing non process-related water use include:

- Wash vehicles used outside the facility less often (Vehicles used inside the facility must be washed after use for safety.) 40 CFR

- Recycle wastewater from vehicle washing. (Your facility may want to evaluate technologies to recycle this wastewater.)

- • Design and maintain landscapes requiring less water

- • Reduce irrigation water use by:
    - Installing timers on sprinkler systems
    - Watering in the early morning or evening when evaporation is lowest
    - Making sure irrigation equipment applies water uniformly
    - Installing drip irrigation systems
    - Using rain sensors.

11.3.2 Techniques for Operational and Housekeeping Changes

The following section describes P2 techniques that pertain to minimizing or eliminating wastes during waste segregation, separation, and preparation processes.

Placing Catch Pans Under Potential Overflows/Leaks   Easy

Placing catch pans or other mini-containment devices near hydraulic lifts, liquid drum storage or dry product storage areas at your food processing facility is an excellent technique to:

(1) Prevent waste from entering drains
(2) Reduce the use of cleanup materials
(3) Reduce wet washing.

While product that hits the ground is generally disposed of as waste or washed down drains, spilled product caught by catch pans can be recycled as animal feed. Catch pans located in a food production area must be cleared regularly and should be removed from the production area for cleaning.

Cover Outside Storage Areas   Easy

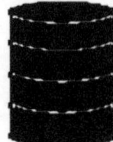

Covering outside storage areas such as waste containers, product storage bins, or cleaning chemical containers is a relatively easy pollution prevention technique that you can implement at your facility. This technique can reduce contaminants in storm water discharges, and help you comply with the Best Management Practice requirements in your facility's storm water permit.

Simple methods of coverage include:
- Moving chemicals inside the plant
- Covering containers with a waterproof tarp when not in use
- Replacing old dumpster covers with new watertight covers
- Replacing or repairing leaking dumpsters.

A pollution prevention technique of moderate expense is to construct an aluminum panel roof under which waste containers, product storage bins, or chemical containers can be stored.

Inspections and Preventive Maintenance of Potential Discharge Areas   Easy

You may find that routine inspections uncover potential problems before they lead to water discharges. Preventive maintenance of outdoor processes can prevent discharges, thus reducing the need for cleanup water and subsequently reducing discharges to storm sewers.

Secondary Containment   Easy - Moderate

Both outdoor and indoor storage facilities should be equipped with secondary containment, which is any device or structure that prevents a spill or leak from reaching the environment. One of the most effective secondary containment methods that can be used in an outdoor storage area is a concrete or asphalt pad surrounded by a berm or curb. The pad and berm prevent any spilled or leaked material from coming in contact with the soil. If a berm is not available, sandbags, or absorbent socks around the perimeter of the area will provide some containment for a short period of time after a spill. Within buildings, depending on construction of the building, the walls and floor provide secondary containment for preventing spills or releases. One of the least expensive secondary containment devices consists of a metal tray covered by a metal grate, which can be used for 55-gallon drums and smaller containers. The container sits on top of the metal grate so that any material or waste that is released from the container simply falls through the grate and is collected by the metal tray underneath. The tray must be large enough to hold the entire volume of the container and should be protected from rainfall.

Monitor Liquid Fill Machines   Easy - Moderate

Ensure liquid fill machines operate correctly and do not overfill cans, jars, etc. Overfill will end up on the floor and be washed down the drain, thereby increasing BOD levels in wastewater. Ways to eliminate overfills include changing the speed at which the machine is operated, adding sensors, and ensuring that the containers flow smoothly through the machine (eliminating jarring movements which could cause spillage).

Cover Outside Drains During Loading and Unloading   Easy

Covering outside drains during loading processes at your food processing facility can prevent spills from reaching storm or process water drains with minimal costs. Preventing spills from entering the wastewater system has several benefits including:

- Preventing potential unauthorized discharges to storm drains
- Preventing high pollutant concentration discharges to treatment plants
- Allowing for a dry precleaning prior to washing a spill area.

Covering Inside Floor Drains
(In Non-production Areas Only)   Easy

As with outside drains, covering floor drains can prevent spills from adding pollutants to wastewater. ***This should be done only in areas where food is not handled.*** Covering drains prevents spills and leaks from flowing directly to the wastewater system. This method provides additional benefits for your facility such as:

- Encouraging dry cleanup by making it more difficult to wash spills down the drains
- Reducing/eliminating non-emergency use and replacement costs of spill equipment (e.g., booms, drying materials) used to prevent spills from reaching floor drains.

Cleaning Prevention   Easy - Difficult

The best way to reduce water use for cleaning spills is to avoid the need for cleaning. Preventing spills from reaching the floor reduces or eliminates water in cleanup. Conducting regular inspections of storage areas and potential spill sites (machines, ovens, conveyors) can prevent spills from occurring, and thus from reaching the wastewater system. Dedicating mixing lines to specific products can reduce changeover cleanups. However, implementation of these lines may be difficult and expensive.

Precleaning and Dry Cleanup   Moderate

For equipment or machinery cleaning, cleaning techniques that reduce water use can save money on water and sewer charges. Techniques such as using squeegees or other dry cleaning equipment prior to wash down, followed by initial rinses with recycled water, have the benefit of allowing you to reduce the time and volume of water in final cleaning. The most important phase of precleaning, however, is dry cleanup.

Dry cleaning is a relatively simple process that involves removing a spill or spent product before washing a surface or container. Many food processing facilities have found dry cleanup to be an easy low cost alternative to hosing spills or unusable product down the drain. They have found that dry cleanup can significantly reduce BOD loading in wastewater discharges, reduce onsite treatment, and reduce the frequency of screen cleaning. When done thoroughly, dry cleanup can prevent all but waste residuals from reaching your facility's wastewater system.

To most effectively conduct dry cleanup, it is important to consider the following:

- All dry wastes should be protected and kept dry to prevent water from contacting the waste, or from entering the drains directly.

- Employees should remove food waste and debris from the production areas and associated equipment with dry methods before using water.

- Solids should be collected from the floor and machines by sweeping and vacuuming into a sanitary container which is kept out of the production area when not is use.

- A stiff broom which is kept sanitized and is cleaned regularly should be used to sweep materials off the floor; scraping and then brushing may be the only effective way to recover some materials from under equipment.

- To allow reuse, clean and store dry cleanup tools and utensils separate from regular wet cleanup gear and in a manner that will not jeopardize the safety of your product.
- Water hoses should be used only as a final alternative to a cleaning task. Any materials on the floor should be removed prior to hose down or wash down.

Dry cleaning can also be used as part of regular washdowns. When emptying cooking ovens or tanks, wastewater pollutants can be reduced by:

- Emptying waste products into barrels instead of pumping down the drain

- Emptying machines by hand rather than hoses.

Skim Grease Traps Regularly   Easy

Skimming grease traps regularly reduces the amount of contaminants entering wastewater. Many facilities hire outside contractors to remove contaminants from grease traps on a periodic basis. For most effective use, however, more frequent skimming as part of regular housekeeping not only ensures that discharged wastewater has reduced contaminants, but also improves your ability to recycle and reuse process waters before discharge.

Screening   Moderate

Placing screens in all drains is an easy pollution prevention technique to collect and prevent solids from entering the wastewater stream. Screening can reduce BOD and solids levels in wastewater, reducing loads on your treatment plant. However, screening should be done only with food safety in mind. Screens need to be cleaned regularly to prevent residue buildup and must be removed from the production area to be cleaned. Cleaning must be done in a sanitary fashion.

Minimizing Pests   Easy - Moderate

When you generate waste, your operating costs increase since you must pay for items, such as hazardous waste disposal which includes waste pesticide, herbicide, and insecticide disposal. By reducing these wastestreams, you can cut the cost of operating your facility. In order to maintain compliance with waste pesticide disposal regulations, you and your staff must conduct a great number of environmental management activities. Instead, your facility can reduce the amount of pesticides, herbicides, and insecticides used at your facility "by design."

Pest prevention by design is the engineering science which will help reduce the need for chemical control of rodents, insects, birds and other vermin. This involves landscape design, building design or remodeling and equipment layout and design. For example, short grass, neatly trimmed shrubs, paved access ways and proper drainage reduce or eliminate shelter areas for pests. Rodents are further discouraged by surrounding the building foundation with an 18 to 24 inch strip of 1/8 inch pebbled rock in a trench approximately 4 inches deep. This makes an excellent area for traps and bait stations.

Other pest control P2 techniques include:

- Eliminating improperly stored equipment, litter, waste, refuse, and uncut weeds or grass within the immediate vicinity of plant buildings or structures to reduce pest harborages.
- Properly sloping, and adequately draining the grounds to avoid contamination of food products through seepage or foot-borne filth. Poor drainage provides a breeding place for insects and microorganisms.
- Positioning outside lighting and focusing it away from buildings to attract night flying insects away from doors and windows.
- Reducing potential bird harborages by screening off harborage areas.
- Eliminating food that may accumulate near malfunctioning exhaust systems.

- Considering various types of rodent, insect, and bird traps. Trapped protected birds must be released.
- Maintaining adequate housekeeping programs.

11.3.3 Techniques for Recycling/Reuse

While reducing the input materials to your food processing operations is the most effective means of pollution prevention, recycling/reusing materials in your operations can be an equally effective way of reducing your solid wastestream. Try using returnable materials containers *(except for food contact materials)* and returnable plastic or wood pallets. Check with your suppliers for other suggestions on how you can recycle/reuse materials that end up in your trash bin. Examples of pollution prevention that involve reduction in waste cleanup that also could be considered methods of recycling are summarized below.

Countercurrent Washes   Moderate

Commonly used in food processing, countercurrent washing can replace parallel tank systems. Countercurrent systems are multistage (tank) systems in which water gets reused in preceding steps. In a three-stage countercurrent wash system, water from the third (final) stage is reused as make-up water for the second stage. Clean water is then used to replenish the final stage. Water from the second stage is reused as make-up water for the first stage. Water from the first stage, which is the dirtiest, is commonly discarded. The countercurrent washing system requires more space and equipment. The benefits of this technique are that it reduces the volume of fresh water used and reduces the volume of wastewater generated. Compared to a non-countercurrent rinse system, this method can reduce water usage by over 50%.

Process Water Reuse   Easy-moderate

Your food processing facility can reuse process water in several applications without compromising food safety. Be sure you comply with all FDA and USDA regulations regarding water reuse. Generally process water that has not been filtered or treated can be used as a first rinse in wash cycles, or for primary cleaning of floors and gutters.

Examples of potential sources of water to be reused include final rinses from tank cleaning; refrigeration defrost; cooler effluent, and sterilizer effluent. Potential opportunities for water reuse include as boiler makeup and caustic dilution.

Water Recirculation Units   Moderate

Water recirculation units can be installed to reuse food processing wash water. The benefits of this technology are that it reduces fresh water use, wastewater discharge, and energy consumption. "Off-the-shelf" units (1) reduce fresh water use because wash bath water is reclaimed and reused and (2) reduce energy use associated with heating the washwater baths. Reclaimed water is already warm so less energy is required to heat it to the required

temperature. Filters from the water recirculation units require disposal and are generally considered nonhazardous solid wastes. ***Food companies which contemplate installing water recirculation units should consult and comply with all appropriate FDA/USDA regulations concerning such a practice.***
The costs associated with installing water recirculation units vary between food processing facilities. Capital expenditures are required for:

- • The water recirculation unit (a minimum of $500);
- • Replumbing of the washwater bath system; and
- • Ongoing operation and maintenance.

Water Used to Chill Products   Moderate

When recycling and recirculating water used to chill products, ***it is important that the water meets FDA and/or USDA standards***. The FDA specifies that any water that contacts foods or food-contact surfaces shall be safe and of adequate sanitary quality. This standard applies to non-meat and non-poultry processing operations and allows for water to be recycled. (Water is recycled through a product cooler, which contains either a refrigerated chiller or a cooling tower to continuously cool the water between cycles.) For these operations, cooling water can be used for initial product washing; however, final washing must be conducted with potable water.

USDA is responsible for meat and poultry processing operations, and has identified three acceptable processes for prechiller water recycling:

- •• Ozonation in a countercurrent flow contact column
- •• Screening, ozonation, sand filtration, and ozonation
- •• Screening, diatomaceous earth (DE) filtration and ozonation.

You will find that any of these processes can significantly improve water quality, reducing solids from between 28% (ozonation alone) and 65% (screen and filtration processes), and COD between 38% and 87%. In addition, these processes have reduced microbial loads by more than 99.9%.

An example of the significant savings you can achieve by recirculating chiller water is provided by the North Carolina Agricultural Extension Service.

If a food processing facility uses 120,000 gallons of water daily to chill its products, it could save 96,000 gallons daily by reconditioning 80% of its waste chiller water. At $1.90 per thousand gallons for water and sewer charges, this plant could save more than 24,000,000 gallons of water valued at more than $45,000 per year. In addition, COD and TSS loads in the effluents could also be reduced by approximately 200,000 lb/yr (assuming an initial average of 1,000 mg/L of COD and TSS, respectively, in the untreated chiller water).

If the surcharge on excess COD is $0.20/lb, the surcharge savings could be almost $40,000 per year. Thus the potential savings for water, sewer, *and surcharges could total as much as $85,000 per year. Other savings might be realized though by-product recovery and reductions in energy costs.[2]*

Residuals Management   Easy - Moderate

Residues are defined as solid by-products that have some positive value or represent no cost for disposal. Food processing residues typically have nutrient/organic matter content that makes them economically recyclable. Some of the more recent technologies for reclaiming by-products for utilization include (1) recovering by-products for use in human food; (2) recovering by-products for animal feeds; (3) use as fertilizers for crop production; and (4) recovery for energy generation.

Recycling Refrigerants   Moderate

If refrigerants are recycled or reclaimed, they are not considered hazardous under federal law. As a food processing facility, recycling or reclaiming your refrigerants will reduce your hazardous waste disposal costs. If you have not done so already, it is important that you consider recycling your refrigerant or contracting a service to reclaim used refrigerant. To assist owners of commercial refrigeration, EPA has published a series of short fact sheets that outline regulations and pollution prevention techniques. For further information, call the Stratospheric Ozone Hotline at 1-800-296-1996.

Reducing/Recycling/Reusing Packaging   Easy

Many businesses across the U.S. generate extremely large amounts of nonhazardous solid waste daily. Much of the waste is from product packaging (e.g., plastic, cardboard, and aluminum). Incinerators and landfills, most often, are the final destination for most of this waste. There are, however, many avenues for diverting the solid waste from a solid waste disposal facility. Inefficiently managed solid waste can lead to excessive and unnecessary expenses for your facility. The following list provides several suggestions and resources to help you better handle your facility's solid waste.

- **Reduce Materials Used.** You can reduce or eliminate a number of input materials to reduce solid wastes generated by your facility. These materials include excess cardboard and plastic packaging.

- **Reuse Materials.** While reducing the input materials to your packaging process is the most effective means of pollution prevention, reusing materials in your operations can be an equally effective way of reducing your solid waste stream. Using returnable materials such as plastic crates or wooden pallets will reduce the amount of waste that ends up in the trash. *Use of returnable materials for food contact should be avoided.*

---

[2] North Carolina Agricultural Cooperative Extension Service. "Bank or Drain: Cut Waste to Reduce Surcharges for Your Dairy Plant." North Carolina Pollution Prevention Pays Program. CD-26. March 1996 (JWM). **http:\\www.bae.ncsu.edu/baeprograms/extension/publicat/wqwm/cd26.html**

- **Recycle Scrap.** Many materials in the packaging process can be recycled, which will prevent them from ending up in the local landfill. They include paper, empty containers, cardboard, pallets, glass, and aluminum. Consult your vendors or local recycling companies for more ideas.

11.3.4 Techniques for Material Substitution and Elimination

As a food processing facility, you should research materials that are safe for the environment (without compromising the safety and quality of your product) and cost less (e.g., by weight or usage amount) that you can use in food processing operations. If it is determined that a material is not needed for a process, eliminate its usage to reduce extra costs in production. By educating all employees about the costs of waste disposal and the benefits of reduction, you can maximize cost savings by implementing pollution prevention techniques throughout your facility.

Laboratory Inventory Reduction   Easy

Keeping laboratory materials to a minimum can benefit your facility by reducing accumulation of unusable chemicals and preservatives. It can also provide incentives to minimize use where possible.

General Inventory Control   Easy

Ordering of Materials. Minimize wastes by ordering quantities of materials that match your needs. When ordering input materials, avoid overstocking by ordering according to usage demands. A good unit price is meaningless if the material goes bad on your shelf and you then have to dispose of it. Buy the largest container that allows you to use all of the contents before they go bad. This minimizes solid waste from packaging.

Inventory Control. Chemical containers labels list the shelf life for the material. You should follow these dates and keep inventories using first-in, first-out practices, which will help you reduce the amount of materials with expired shelf lives.

Using Alternative Refrigerants   Moderate

Your facility should consider using alternative refrigerants for your equipment. Many new alternative refrigerants are being marketed for use in stationary refrigeration equipment. You should ask your refrigerant supplier if an alternative is available and whether it is on EPA's Significant New Alternatives Policy (SNAP) program list. EPA's SNAP program determines what risks alternatives to refrigerants pose to human health and the environment. EPA evaluates the alternative refrigerant's ozone-depleting potential, global warming potential, flammability, and toxicity. The SNAP evaluation, however, does not determine whether the alternative will provide adequate performance or will be compatible with the components of a refrigeration system. *Food processors should consult with their refrigeration supplier/engineer prior to considering a SNAP refrigerant to ensure that safe temperature parameters for their product will not be compromised.* To assist owners of commercial refrigeration, EPA has published a series of short fact sheets that outline regulations and pollution prevention techniques. For further information, call the Stratospheric Ozone Hotline at 1-800-296-1996.

11.4   Voluntary Programs

Over the last several years, an important change has been taking place in EPA's national strategy for protecting the environment. Through an array of partnership programs that EPA collectively refer to as *Partners for the Environment*, EPA is demonstrating that voluntary goals and commitments achieve real environmental results in a timely and cost-effective way. In addition to traditional approaches to environmental protection, EPA is building cooperative partnerships with a variety of groups, including small and large businesses, citizen groups, state and local governments, universities and trade associations.

The results of the *Partners for the Environment* effort are impressive. Thousands of organizations are working cooperatively with EPA to set and reach environmental goals such as conserving water and energy, and reducing greenhouse gases, toxic emissions, solid wastes, indoor air pollution and pesticide risk. EPA's partners are making pollution prevention a central consideration in doing business. Partnership also means that EPA is working cooperatively with the private sector to provide stakeholders with effective tools to address environmental issues. And these partners are achieving measurable environmental results often more quickly and with lower costs than would be the case with regulatory approaches. EPA views these partnership efforts as key to the future success of environmental protection.

EPA's voluntary pollution prevention programs, such as the Environmental Leadership Program (ELP), Project XL, and WasteWi$e, are designed to promote industrial environmental excellence. Some programs offer opportunities for both trade association and individual companies to participate. As of 1996, trade associations representing the food processing industry and/or individual companies were participating in most of these voluntary programs. Several federally sponsored demonstration programs (e.g., Climate Wise, Green Lights, and NICE[3]) focus on energy savings in industrial operations. Although energy use is not regulated, energy conservation and pollution prevention are interrelated. As of 1996, a small number of food processing companies were participating in these programs.

EPA has produced a reference guide that describes 38 of its voluntary pollution prevention programs, entitled "Partnerships in Prevention Pollution: A Catalogue of the Agency's Partnership Programs" (1996). This document can be accessed at **http://www.epa.gov/ partners/**.

11.4.1   EPA Programs

Environmental Leadership Program

From 1994 to 1996, EPA's Office of Compliance tested a national initiative, the Environmental Leadership Program (ELP), with twelve industrial facilities (e.g. printing, waste management services, etc.) and federal installations. Note: No food processing facilities participated in the pilot phase of this initiative.

The program provided recognition and certain other benefits to facilities that demonstrated strong commitments to continued compliance and "beyond compliance" efforts. Two of the criteria for participation were that the facility had to have a good record of compliance with environmental laws, regulations and permits, and the facility had to demonstrate it had an environmental management system (EMS) that met ELP requirements. EPA is reviewing the ELP's results before further action on this program. For additional information, visit the ELP Home Page at **http://es.epa.gov/elp/.**

Project XL

Project XL was initiated in March 1995 as a part of President Clinton's *Reinventing Environmental Regulation* initiative. Project XL, which stands for "e**X**cellence and **L**eadership," is a national initiative that tests innovative ways of achieving better and more cost-effective public health and environmental protection. The information and lessons learned from Project XL will be used to assist EPA in redesigning its current regulatory and policy-setting approaches. Project XL encourages testing of cleaner, cheaper, and smarter ways to attain environmental results superior to those achieved under current regulations and policies, in conjunction with greater accountability to stakeholders.

EPA and program participants will negotiate and sign a Final Project Agreement, detailing specific objectives that the participant (regulated entity) shall satisfy. In exchange, EPA will allow the participant a certain degree of regulatory flexibility and may seek changes in underlying regulations or statutes. Participants are encouraged to seek stakeholder support from local governments, businesses, and environmental groups. EPA hopes to implement fifty pilot projects in four categories including facilities, sectors, communities, and government agencies regulated by EPA. Applications will be accepted on a rolling basis and projects will move to implementation within six months of their selection.

JACK M. BERRY INC. Jack M. Berry Inc. is a mid-sized juice-processing facility in LaBelle, Florida.

Innovative Approach: Jack M. Berry Inc. is developing a facility-wide comprehensive operating plan that consolidates environmental permits and all operating procedures into a single manual for the facility. The project builds in stakeholder participation, and will be evaluated with appropriate public notices every five years. The project may be consolidating seven Federal, State, and local environmental permits by developing and gaining approval for just one comprehensive operating permit instead of many each year. It is also improving compliance with environmental requirements by involving staff in the development of the facility-wide operating plan and by using simple language to describe more clearly what is required by law.

Benefits for the Environment: In the first year of the project, the facility eliminated several hazardous wastestreams, and an 88-acre area previously used to disperse wastewater, which relieved the community of irritating odor problems. The facility is also expected to: (1) reduce air emissions of volatile organic compounds, sulfur dioxide, and nitrogen oxides; and (2) further reduce the number and types of solvents and lubricants used onsite and replace them with a number of environmentally-friendly materials.

Benefits to the Facility: Jack M. Berry Inc. will save significant expenditures by eliminating the costly requirement of preparing multiple permit applications every few years. This results in reduced lender concern about future operational status, which, in turn, can translate into lower interest rates for long-term loans. In addition, as a result of audits during the project's first year, the company's new work procedures are expected to result in 50 percent savings in environmental control investments, improved worker safety, and substantially reduced employee training costs.

Stakeholder Involvement: Jack M. Berry Inc. has been working to ensure that those parties with a stake in the environmental concepts of its project are informed and have had an opportunity to participate in the development of the project.

As of March 1998, more than 50 proposals have been reviewed to date. Seven pilot projects, including Jack M. Berry Inc. of Labelle, Florida (see box below), have signed final project agreements and are being implemented, and twenty proposals are in the development stage. More information on the Jack M. Berry pilot project can be found at **http://yosemite.epa.gov/xl/xl_home.nsf/all/berry.html/**.

For additional information on Project XL, including application procedures and criteria, see the April 23, 1997 Federal Register Notice, call the Project XL Information Line at (703) 934-3239, or use the Project XL fax-on-demand line at (202) 260-8590.   Additional information can be obtained from EPA's fact sheet entitled, "What Is Project XL? Excellence and Leadership in Environmental Protection" (EPA 231-F-97-001), March 1998, and other project-specific fact sheets, all of which are available on the Internet at **http://yosemite.epa.gov/xl/xl_home.nsf/all/homepage/** or via Project XL's fax-on-demand line.

WasteWi$e Program

The WasteWi$e Program was started in 1994 by EPA's Office of Solid Waste and Emergency Response. The program is aimed at reducing municipal solid wastes by promoting waste minimization, recycling collection, and the manufacturing and purchase of

---

recycled products. As of January 1998, the program had about 700 partners spanning more than 35 industry sectors. Partners include large corporations, as well as small and medium-sized businesses. WasteWi$e has 59 endorsers, mainly membership-based organizations, from more than 15 industry sectors. Partners agree to identify and implement actions to reduce their solid wastes and must provide EPA with their waste reduction goals along with yearly progress reports. EPA, in turn, provides technical assistance to partner companies and allows the use of the WasteWi$e logo for promotional purposes. For more information, contact the WasteWi$e Hotline at 800-EPA-WISE (372-9473) or access the WasteWi$e Home Page via the Internet at **http://www.epa.gov/epaoswer/non-hw/reduce/wstewise**.

Climate Wise

Climate Wise, a unique, government-industry partnership jointly sponsored by the U.S. Department of Energy (DOE) and EPA, helps businesses turn energy efficiency and environmental performance into a corporate asset. Climate Wise, a voluntary program, was designed to help the United States honor its international commitment to reducing greenhouse gas emissions to 1990 levels by the year 2000. Climate change prevention measures can continue to be a prime focus of international negotiations in the future.

Companies participating in Climate Wise are finding that improving energy efficiency and reducing greenhouse gas emissions save them money and boost productivity. Climate Wise Companies already expect to save more than $300 million by the year 2000. Becoming a partner is easy. To join, companies must complete a one-page partnership agreement; submit a Climate Wise Action Plan within six months that identifies specific cost-effective energy efficiency and pollution prevention measures; and report results annually while striving for continuous improvement. In return, participants in the Climate Wise program receive DOE and EPA help in identifying actions that both save energy and reduce costs. For example, Climate Wise partners receive an innovative action plan development software program that provides more than 50 case studies, a list of proven energy efficiency technologies, and tools to quantify the results of their actions. Also, Climate Wise companies can receive access to free pollution prevention and energy efficiency assessments. In addition, companies receive public recognition for their efforts.

Over 300 current partners have taken advantage of the program's many service offerings, including financial information sources, supporting documents, and peer exchange opportunities. For more information, call 202-260-4407 or access the ClimateWi$e Home Page via the Internet at **http://www.epa.gov/climatewise/**.

Green Lights Program

Green Lights is an innovative, voluntary pollution prevention program sponsored by EPA. The primary purpose of the Green Lights Program is to encourage U.S. organizations to install energy-efficient lighting, in order to prevent the creation of air pollution (including greenhouse gases, acid rain emissions, air toxics, and tropospheric ozone), solid waste, and other environmental impacts of electricity generation. As of April 1998, the program had over 2,500 members which included major corporations; small and medium sized businesses; federal, state and local governments; non-profit groups; schools; universities; and health care facilities.

By joining Green Lights, partners agree to install energy efficient lighting where profitable as long as lighting quality is maintained or improved. EPA agrees that your commitment to survey buildings and complete lighting upgrades is contingent upon the availability of appropriated funds or third-party financing resources. EPA provides technical assistance to the participants through a decision support software package, workshops and manuals, and a financing registry. EPA's Office of Air and Radiation is responsible for operating the Green Lights Program. For additional information, contact Green Light/Energy Star Hotline at 202-775-6650 or call toll-free at (888) STAR-YES (782-7937)]. Information can also be accessed using the fax-back system at 202-564-9659 or by accessing the Green Lights Home Page via the Internet at **http://www.epa.gov/greenlights.html/**.

NICE[3]

The U.S. Department of Energy (DOE) and EPA's Office of Pollution Prevention are jointly administering a grant program called The National Industrial Competitiveness through Energy, Environment, and Economics (NICE[3]). By providing grants of up to 50 percent of the total project cost, the program encourages industry to reduce industrial waste at its source and become more energy-efficient and cost-competitive through waste minimization efforts. Grants are used by industry to design, test, demonstrate, and assess the feasibility of new processes and/or equipment with the potential to reduce pollution and increase energy efficiency. The program is open to all industries; however, priority is given to proposals from participants in the pulp and paper, chemicals, primary metals, and petroleum and coal products sectors. For more information, contact DOE's Golden Field Office at 303-275-4729 or access **http://www.oit.doe.gov/Access/nice3/basicbody.html/**.

11.4.2 Trade Association/Industry Programs

Trade associations and other industry-related groups are developing programs that promote pollution prevention opportunities. The following are examples of these programs developed for the food processing industry.

Food Manufacturing Coalition for Innovation and Technology Transfer

Initiated on January 23, 1996, the Food Manufacturing Coalition (FMC) is an ongoing, industry-driven technology transfer program. The objectives of the FMC are to (1) improve the food manufacturing industry's productivity and environmental quality through technological innovation and commercialization, and (2) address and solve high priority, industry-wide environmental problems. The program is open to companies of all sizes.

Members of the FMC discussed and selected specific high priority areas initially identified through surveys conducted by 8 national trade associations partnering in the project. A total of 20 potential projects directed toward maximizing air and water quality, minimizing sold waste and toward increased control and processing efficiencies were designated for further analysis and effort. These topic areas were further refined into detailed needs statements that are being broadly disseminated to the research and development community asking for technical ideas and interest in joint efforts. The needs and suggested technological approaches will result in State-of-the Art reports that document alternative technologies available for follow-up in the form of co-development, licensing, Small Business Innovation Research Grants, or other strategies leading to potential commercialization.

For more information on the FMC program, contact R.J. Phillips & Associates, Inc. at (703) 406-0072 or send e-mail to rphil1140@aol.com. Additional information can also be obtained by accessing the FMC webpage at **http://ceres.esusda.gov/fmc/**.

Communicating CAA Section 112 (r) Risk Management Program Requirements

The Food Industry Environmental Council (FIEC) a coalition of more than 50 food processors and trade associations, has developed materials to assist food processors in communicating with the public about risk management programs covered under the CAA Section 112(r). These communication materials include the following:

- "Backgrounders" on ammonia, chlorine and propane;
- A computer disk with the shell of a tri-fold brochure and filler language;
- Communication guidelines;
- A question and answer document; and
- A resource and reference document.

The communication packages are available from your food trade association.

APPENDIX A
SUMMARY OF MAJOR REGULATIONS
FROM THE CFR

The following contains brief summaries of the major environmental regulations from the *Code of Federal Regulations* (CFR) that are applicable to food processors. Table A-1 shows the regulations and the CFR citations that are presented in Appendices A.1-A.6. These summaries can assist you in identifying specific regulatory requirements. Appendix A.7 lists pending and proposed regulations. You should note that these materials are intended solely as guidance. Because applicable regulations are specific to each individual facility, you should use the *Federal Register* or the CFR to determine your facility's requirements.

Table A-1. Regulations and CFR Citations Presented in Appendix A

Appendix	Statute	Regulation	
A.1	CWA	NPDES Permit Program	40 CFR 122
		Pretreatment Regulations	40 CFR 403
		Discharge of Oil	40 CFR 110
		Oil Pollution Prevention	40 CFR 112
		Designation of Hazardous Substances	40 CFR 116
		Determination of RQs for Hazardous Substances	40 CFR 117
A.2	SDWA	National Primary Drinking Water Regulation	40 CFR 141
		National Secondary Drinking Water Regulation	40 CFR 143
		Underground Injection Control Program	40 CFR 144
A.3	CAA	Subpart M National Emission Standards for Asbestos	40 CFR 61
		Chemical Accident Prevention Provisions	40 CFR 68
		Protection of Stratospheric Ozone	40 CFR 82
A.4	EPCRA	Emergency Planning and Notification	40 CFR 355
		Hazardous Chemical Reporting	40 CFR 370
		Toxic Chemical Release Reporting	40 CFR 372

Table A-1. Regulations and CFR Citations Presented in Appendix A

Appendix	Statute	Regulation	
A.5	CERCLA	Designation, Reportable Quantities and Notification	40 CFR 302
A.6	RCRA	Generator Classifications and Requirements	40 CFR 261.5 & 262.34
		Hazardous Waste Generator Requirements	40 CFR 262
		Hazardous Waste Transporter Requirements	40 CFR 263
		Land Disposal Restrictions	40 CFR 268
		Underground Storage Tanks (USTs)	40 CFR 280
	FIFRA	Not summarized in this document.	
	TSCA	Not summarized in this document.	

APPENDIX A. 1
SUMMARY OF PRINCIPAL REGULATIONS
UNDER THE CLEAN WATER ACT

The following section provides a summary of the principal regulations developed pursuant to the CWA that are applicable to the food processing industry. The regulations included are:

- **40 CFR 122** - NPDES Permit Program
- **40 CFR 403** - Pretreatment Regulations
- **40 CFR 110** - Discharge of Oil
- **40 CFR 112** - Oil Pollution Prevention
- **40 CFR 116** - Designation of Hazardous Substances
- **40 CFR 117** - Determination of Reportable Quantities for Hazardous Substances

40 CFR 122
EPA Administered Permit Programs, The National Pollutant Discharge Elimination System

Definition of a Point Source (40 CFR 122.2): For the purposes of the CWA, **point source** means any discernible, confined, and discrete conveyance, including but not limited to, any pipe, ditch, channel, tunnel, conduit, well, discrete fissure, container, rolling stock, concentrated animal feeding operation, landfill, leachate collection system, vessel or other floating craft from which pollutants are or may be discharged. This term does not include return flows from irrigated agriculture or agriculture storm water runoff.

40 CFR 122 - DIRECT DISCHARGES	
Requirements	**Compliance Dates**
Discharge Limitations: • Effluent limitations contained in the NPDES permit. **Monitoring and Reporting Requirements:** Note: All direct dischargers are required to obtain an NPDES permit. The NPDES permit outlines the discharger's specific monitoring and reporting requirements.	Compliance with specific permit limitations upon effective date of the permit.

40 CFR 122 - DIRECT DISCHARGES	
Requirements	**Compliance Dates**
• <u>Permit Applications</u> - containing the information required under 122.21(f), (g), and (k) (application requirements for new sources and new discharges)	Permit applications are to be submitted 180 days prior to the commencement of discharge. Applications for permit renewal are required to be submitted 180 days before the existing permit expires.
• <u>Planned Changes</u> - notification to the Director as soon as possible of any planned physical alteration or addition that meets the criteria in 122.41(l)(1)	As soon as possible, when applicable
• <u>Anticipated Noncompliance</u> - advance notification to the Director of any planned changes that may result in permit noncompliance	In advance of changes, when needed
• <u>Monitoring Reports</u> - monitoring results must be submitted as required by the NPDES permit (at least annually). All monitoring must be conducted using 40 CFR 136 methods.	At least annually or more frequently as required by permit
• <u>Compliance Schedules</u> - reports of compliance or noncompliance with compliance schedule requirements	Within 14 days of each compliance schedule date
• <u>24-Hour Reporting</u> - of any noncompliance that may endanger health or the environment, including the information listed in 122.41(l)(6)	Within 24 hours
• <u>Anticipated and Unanticipated Bypass</u> - notification as required under 122.41(m)	At least 10 days prior to anticipated bypass. Within 24 hours of unanticipated bypass
• <u>Discharge of Toxic Pollutants</u> - notification to the Director of activity that results in the discharge of toxic pollutants not limited in the permit, if it exceeds the levels outlined in 122.42(a)(1)	As soon as facility knows or has reason to believe that levels will be exceeded
• <u>Storm Water Permit Applications</u> - submission of either individual permit application or general permit applications	
- Individual permit applications must include the information in 122.26(c)(1)	Individual permit applications for existing facilities were due October 1, 1992. New facilities must submit an application 180 days prior to commencement of industrial activity

40 CFR 122 - Direct Discharges	
Requirements	**Compliance Dates**
- Facilities to be covered under a baseline general permit must file a Notice of Intent (NOI)	NOIs from existing facilities were due prior to October 1, 1992. NOIs were due prior to September 9, 1997 in order to be covered under the administratively-extended baseline general permit.
– Facilities to be covered under a multi-sector general permit must file a NOI.	Deadlines for NOIs differed for facilities in operation prior to September 29, 1995, and those who commenced operations after September 29, 1995.
• <u>Other Storm Water Reports</u> - submission of other reports as required under a facility's storm water discharge permit. These reports may include pollution prevention plans and monitoring reports. **Recordkeeping Requirements:** • Records of monitoring information as required under 122.41(j) must be kept for at least three years.	Due dates as required by permits

40 CFR 403
General Pretreatment Regulations for Existing and New Sources of Pollution

40 CFR 403 - Indirect Discharges	
Requirements	**Compliance Dates**
Discharge limitations: • Prohibited discharge standards (general and specific) in 40 CFR 403.5 • Applicable local limits	
Monitoring and Reporting Requirements: Note: Reports must be submitted whether or not the facility has been issued a permit.	

40 CFR 403 - INDIRECT DISCHARGES	
• <u>Baseline Monitoring Reports (BMR)</u> - containing the information required under 40 CFR 403.12(b).	BMRs from existing sources are due within 180 days after the effective date of a categorical pretreatment standard. BMRs from new sources are due 90 days prior to commencement of discharge.
• <u>Compliance Schedule Progress Reports</u> - containing the information required under 40 CFR 403.12(c)(3)	Due within 14 days of completing compliance schedule milestone or due date
• <u>90-Day Compliance Report</u> - containing the information required under 40 CFR 403.12(d)	Due within 90 days following date for final compliance or for new sources, following the commencement of introduction of wastewater to the POTW
• <u>Periodic Reports on Continued Compliance</u> - containing the information in 40 CFR 403.12(e) (including monitoring data for all categorically regulated pollutants). All monitoring must be conducted using 40 CFR 136 methods.	Must be submitted at least semiannually
• <u>Notice of Potential Problems Including Slug Loadings</u>	Immediately to the Control Authority upon identification of discharges that could cause problems to the POTW
• <u>Notice of Changed Discharge</u> - advanced notification of any substantial change in the volume or character of pollutants in the discharge (including hazardous wastes)	Prompt notification in advance of any substantial change
• <u>Notice of Violations and Resampling</u> - notification of violation and results of sampling	Notice within 24 hours, results of resampling within 30 days
• <u>Notification of Hazardous Waste Discharge</u> - notification to the POTW, EPA, and the state of the hazardous wastes discharged to the POTW	One time notification, unless changes to discharge
Recordkeeping Requirements: • Monitoring records including the information listed in 403.12(o) must be maintained for at least 3 years	

40 CFR 110
Discharge of Oil

Applicability:

Prohibited discharges include certain discharges to U.S. navigable water, adjoining shorelines, or to waters of the contiguous zone, occurring in connection with activities under the Outer Continental Shelf Lands Act of the Deepwater Port Act, or those that may affect U.S. natural resources.

May be applicable to food processing facilities using oil and that are either located by a municipal storm sewer that discharges to waters or near streams or bodies of water.

40 CFR 110	
Requirements	

• Discharge of oil is prohibited that:
 	
  - Violates applicable water quality standards, or
 	
  - Causes a film or sheen upon or discoloration of the surface of the water or adjoining shorelines or causes a sludge or emulsion to be deposited beneath the surface of the water or upon the adjoining shorelines	

• Notification must be provided immediately to the National Response Center at 1-800-424-8802 or 202-426-2675 in the Washington, DC metropolitan area of any discharge of oil in violation of the Section 311(b)(3).	Immediately

40 CFR 112
Oil Pollution Prevention

Applicability:

Non-transportation related onshore and off-shore facilities engaged in drilling, producing, gathering, storing, processing, refining, transferring, distributing, or consuming oil and oil products that could reasonably discharge oil in harmful quantities, as defined in 40 CFR 110.

Note: 40 CFR 112 does not apply to facilities, equipment, or operations which are not subject to EPA jurisdiction.

40 CFR 112	
Requirements	**Compliance Dates**
Prepare and implement Spill Prevention Control and Countermeasure plans meeting the requirements of 40 CFR 112.3 - 112.7	Existing sources: Plans in effect New sources: Prepare plan within 6 months of beginning operation and fully implement in no later than 1 year
• Submit report as described in 40 CFR 112.4 when discharged oil > 1,000 gallons in single spill event or discharged oil in harmful quantities in two spill events	Within 60 days of becoming subject to reporting requirements
• Review, evaluate, and update plan as required under 40 CFR 112.5	Review plan once every 3 years, amend plan within 6 months, if needed
• Submit facility response plan as described in 40 CFR 112.20 and develop and implement facility response training and drill exercise as described in 40 CFR 112.21	Existing sources: as described in 40 CFR 112.20 New source: Prior to start of operations

Additional requirements not addressed here include facility response plans, preparedness drills and exercises, and training.

40 CFR 116 and 117
Designation of Hazardous Substance and 40 CFR 117 Determination of Reportable Quantities for Hazardous Substances

Applicability:

40 CFR 117 does not apply to facilities that discharge the substance under an NPDES permit or to a POTW, as long as any applicable effluent limit or pretreatment standard is met.

Requirements:

40 CFR 116.4 designates hazardous substances and 40 CFR 117.3 establishes the Reportable Quantity (RQ) for each substance listed in 40 CFR 116. When an amount equal to or in excess of the RQ is discharged, the facility must provide notice to the Federal government following DOT requirements in 33 CFR 153.203.

APPENDIX A.2
SUMMARY OF PRINCIPAL REGULATIONS
UNDER THE SAFE DRINKING WATER ACT

The following section provides a summary of the principal regulations developed pursuant to the SDWA that may apply to the food processing industry:

- **40 CFR 141** - National Primary Drinking Water Regulations
- **40 CFR 143** - National Secondary Drinking Water Regulations
- **40 CFR 144** - Underground Injection Control Program.

40 CFR 141
National Primary Drinking Water Regulations

Applicable Subparts:

Public water systems (PWSs) applicable to food processing facilities includes:

- Community water system - A PWS which serves at least 15 service connections used by year round residents or regularly serves at least 25 year-round residents.
- Non-transient non-community water system - A PWS that is not a community water system and that regularly serves at least 25 of the same persons for over 6 months of the year.

40 CFR 141	
Requirements	
Maximum Containment Levels - Subparts B, G	All Regulations in effect
Maximum Containment Level Goals - Subpart F	
Monitoring and Analytical Requirements - Subparts C, H, I	
Reporting, Public Notification and Recordkeeping - Subparts D, H, I	
Surface Water Treatment Rule - Subpart H	Under development

Recordkeeping Requirements (40 CFR 141.33, 144.75, 141.80 and 141.91)

Records Pertaining to	
Bacteriological analyses	At least 5 years
Chemical analyses	At least 10 years
Actions taken to correct violations	At least 3 years after last action taken
Sanitary survey reports	At least 10 years
Variances or exemptions	At least 5 years following expiration
Lead and copper control	At least 12 years

Lab Reports Summary Requirements (40 CFR 141.33, 141.74 and 141.80)

Sampling Information	
Date, place, and time of sampling	Date of analysis
Name of sample collector	Laboratory conducting analysis
Identification of sample: • Routine or check sample • Raw or treated water	Name of person responsible for analysis Analytical method used Analysis results

Reporting Requirements for Check Sampling

Contaminant	
Microbiological	Must report to state within 48 hours when any check sample confirms the presence of coliform bacteria.
Nitrate	Must report to state within 24 hours if check sampling confirms MCL has been exceeded
All others	Must be reported to the state within 10 days after the end of the month in which the sample was received.

MCL Violations

Contaminant	Violation
Inorganic chemicals (expect nitrate) and organic chemicals (except THMs)	If average of results from initial sample plus 3 check samples exceeds MCL
Nitrate	If average of results from initial sample plus the check sample exceeds MCL
THMs	If average of results from present quarter plus those of 3 preceding quarters exceeds MCL*
Radionuclides (natural and man-made)	If average annual concentration exceeds MCL**
Microbiological (coliform testing): membrane filter and multiple-tube fermentation	If any of the MCLs are exceeded

*   Quarter means a 3-month period. For convenience, calendar quarters are used.
**   Based on individual analyses of 4 consecutive quarterly samples or a single analysis of an annual composite of 4 quarterly samples.

Public Notification Requirements, 40 CFR 141.32

Violation or Condition	Required Timing			
	72 Hours	14 Days	45 Days	3 Months
Violation of an MCL, acute	3, 4, 5	2, 4, 5	1, 4, 5	1, 4, 5
Violation of an MCL, non-acute		2, 4, 5	1, 4, 5	1, 4, 5
Failure to monitor; failure to follow compliance schedule; or failure to use approved testing procedure				2, 4, 5
System granted a variance or exemption				1, 4, 5

1 - Direct mail   2 - Local newspaper   3 - By local radio and/or TV
4 - Hand delivery   5 - Continuous posting in conspicuous places

40 CFR 143
National Secondary Drinking Water Regulations

Applicable Subparts:

These regulations are not Federally enforceable but are intended as guidelines for States.

40 CFR 143	
Component	**Regulatory Recommendation**
Standards	Secondary MCLs exist for 15 contaminants
Monitoring	Conducted at least as frequently as the monitoring performed for inorganic chemicals listed in the National Interim Primary Drinking Water Regulations and more frequently for parameters such as pH, color, and odor
Analytical Methods	pH, copper, and fluoride should be analyzed consistent with methods described in 40 CFR 141. Other contaminants should be analyzed using the procedures specified in 40 CFR 143.4(b).
Notification	Community water systems that exceed the secondary MCL for fluoride, but do not exceed the primary MCL, should notify (using the public notice provided in 40 CFR 143.5(b)) all billing units annually, all new billing units at the time service begins, and the state public health officer.

40 CFR 144
Underground Injection Control Program

Applicable Subparts:

Well classifications applicable to food processing facilities:

- Class I - Wells used to inject hazardous or nonhazardous wastes beneath the lowermost formation containing, within one-quarter mile of the well-bore, an underground source of drinking water.

- Class V - Injection wells not included in other classes.

40 CFR 144	
Requirements	
Any underground injection is prohibited unless authorized by permit or rule. Construction of any well required to have a permit is prohibited until the permit has been issued.	
Authorization by Rule Requirements:	
Reporting Requirements:	
• Inventory information as specified in 40 CFR 144.26	One year after the date of approval or effective date of the UIC program for the State.
• 24-hour notification of noncompliance that may endanger health or the environment (Class I wells) as required in 40 CFR 144.28(b)	Orally within 24 hours and written five days
• Plugging and abandonment plan (Class I wells) as required in 40 CFR 144.28(c).	One year after the effective date of the UIC program in the State (EPA administered programs)
• Reports containing the information required in 40 CFR 144.28(h)(I) (Class I wells)	Quarterly
• Notice of abandonment as required in 40 CFR 144.28(j)	As specified by the Director
• Plugging and abandonment report as required in 40 CFR 144.28(k)	Existing wells: No later than 4 years from approval or promulgation of UIC program. New wells: Reasonable time before construction is expected to begin
Authorization by Permit Monitoring requirements: • All owners and operators (even those authorized by rule, unless authorized for life of the well) are required to submit a permit application containing the information in 40 CFR 144.31.	

APPENDIX A.3
SUMMARY OF PRINCIPAL REGULATIONS
UNDER THE CLEAN AIR ACT

The following sections provide summaries of some of the principal regulations developed pursuant to the CAA that may apply to the food processing industry. The section includes:

- **40 CFR 61** - Subpart M National Emission Standards for Asbestos
- **40 CFR 68** - Risk Management Planning
- **40 CFR 82** - Protection of Stratospheric Ozone

Additional requirements exist for boilers, NSPS, and NESHAP; see the regulations for more information.

40 CFR 61 - Subpart M
National Emission Standard for Asbestos
Standard for Demolition and Renovation 61.145
Standard for Spraying 61.146
Standard for Insulating Materials 61.148
Standard for Waste Disposal for Manufacturing, Fabricating, Demolition, Renovation, and Spraying Operations 61.150

Applicability:

•• 40 CFR 61.145 is applicable to owners or operators of a demolition or renovation activity

• 40 CFR 61.146 is applicable to owners or operators of an operation in which asbestos-containing materials are spray applied.

Affected Processes:

- For demolition, requirements in 40 CFR 61.145(b) and (c) apply if the combined amount of Regulated Asbestos-Containing Material (RACM) meets criteria listed in 40 CFR 61.145(a)(1)(i) or (ii).

- For renovation, requirements in 40 CFR 61.145(b)(and (c)) apply if the combined amount of RACM to be stripped, removed, dislodged, cut, drilled, or disturbed meets the criteria in 61.145(4)(i) or (ii).

- All RACM must be removed from a facility being demolished or renovated before any activity begins that would break up, dislodge, or disturb the material or preclude access to the material for removal.

- When a facility component that contains, is covered with, or is coated with RACM is being taken out of the facility as a unit or in sections, the procedures in 40 CFR 61.145(c)(2) must be followed; and when RACM is stripped from a facility component while it remains in place at the facility, procedures in 40 CFR 61.145(c)(3) must be met.

- After a facility component covered with, coated, with, or containing RACM is taken out of the facility, it must be handled according to the procedures in 40 CFR 61.145(c)(4). Large components such as reactor vessels, large tanks, and steam generators must be handled according to procedures in 40 CFR 61.145(c)(5).

- • All RACM must be handled according to procedures in 40 CFR 61.145(c)(6).

- No RACM can be stripped, removed, or otherwise handled or disturbed at a facility unless at least one onsite representative is trained in compliance with the regulations.

- Under 40 CFR 61.146, material that contains more than 1% asbestos cannot be used for spray application on buildings, structures, pipes, and conduits.

- Under 40 CFR 61.148, no owner or operator may install or reinstall on a facility component any insulating materials that contain commercial asbestos if the materials are either molded and friable or wet-applied and friable after drying; and this does not apply to spray-applied insulating materials regulated under 40 CFR 61.146.

- Under 40 CFR 61.150, each owner or operator of any source covered under 40 CFR 61.145 or 61.146 must:

  - Discharge no visible emissions to the outside air during the collection, processing, packaging, or transporting of any asbestos-containing waste material generated by the source, or use one of the emission control and waste treatment methods specified in 40 CFR 61.150(a)((1) through (4).

  - Dispose of all asbestos-containing waste material as soon as practical at sites as listed in 40 CFR 61.150(b).

- Mark vehicles used to transport asbestos-containing waste material as in 40 CFR 61.150(c).

Exemptions:

• If the facility is being demolished under State or local government order because the facility is structurally unsound or in danger of imminent collapse, only 40 CFR 61.145(b)(1), (b)(2), b(3)(iii), (b)(4) (except (b)(4)(viii)), (b)(5), and (c)(4) through (c)(9) of this section apply.

• RACM does not need to be removed before demolition if it meets the criteria in 40 CFR 61.145(c)(1)(i), (ii), (iii), or (iv).

• Spray-on application of materials is not subject to 40 CFR 61.146 when the asbestos fibers in the materials are encapsulated with a bituminous or resinous binder during spraying and the materials are not friable after drying.

• Owners and operators of sources subject to 40 CFR 61.146 are exempt from the requirements of 40 CFR 61.05(a), 61.07, and 61.09.

• Requirements in 40 CFR 61.150(a) do not apply to demolition and renovation for Category I nonfriable ACM waste and Category II nonfriable ACM waste that did not become crumbled, pulverized, or reduced to powder.

Reporting and Recordkeeping Requirements

• Owner or operator of demolition or renovation activity must submit and update written notice containing the information in 40 CFR 61.145(b)(4)(i) through (xvii).

• Spray-on application of materials that contain more than 1% asbestos on equipment and machinery are subject to the notification and procedural requirements in 40 CFR 61.146(b)(1) and (2).

• Waste shipment records must be maintained for all asbestos-containing waste as described in 40 CFR 61.150(d).

40 CFR 68
Risk Management Planning

Applicability:

Owners or operators of stationary sources that have more than a threshold quantity of a regulated substance in a process, as determined under 40 CFR 68.115.

Date of Applicability:

The latest of the following dates:

- June 21, 1999
- Three years after the date on which a regulated substance is first listed
- The date on which a regulated substance is first present above a threshold quantity.

Applicable Program:

A covered process is eligible under one of the three following programs. If at any time a covered process no longer meets the eligibility criteria of its Program level, the owner or operator shall comply with the requirements of the new Program level that applies and update the RMP.

Program 1 - For five years prior to submission of the RMP, the process has not had an accidental release of a regulated substance that led to death, injury, or response or restoration activities for exposure of an environmental receptor, and the distance to a toxic or flammable endpoint for a worst-case release assessment is less than the distance to any public receptor, and emergency response procedures have been coordinated between the stationary source and local emergency planning and response organizations.

Program 2 - A covered process not subject to Program 1 or Program 3

Program 3 - A covered process, not subject to Program 1 and either; the process is in SIC code 2611, 2812, 2819, 2821, 2865, 2869, 2873, 2879, or 2911, or, the process is subject to the OSHA process safety management standard 29 CFR 1910.119.

General Requirements:

- Implement a Risk Management Program that includes a hazard assessment, release prevention program, and emergency response program.
- Submit a Risk Management Plan with a registration that includes all covered processes.

Risk Management Plan Requirements:

- An executive summary describing elements of the RMP
- A single registration form covering all regulated substances
- Worst-case release scenario information
- Five-year accident history information
- Alternative release scenarios information (for Program 2 and 3 processes)
- Emergency response program information
- Certification statement
- Regular review and updates to the RMP
- Additional information for Programs 2 and 3.

Other Requirements:

- Maintain records for five years
- Information available to the public
- Additional permit requirements for facilities permitted pursuant to 40 CFR 70 or 71
- Provide access to implementing agency for RMP audits.

Additional Program 1 Requirements:

- Analyze worst-case release scenarios, document public receptor is beyond endpoint, and submit
- Complete five year accident history for the process and submit
- Ensure that response actions coordinated with local agencies
- Certify as specified in 40 CFR 68.12(b)(4).

Additional Program 2 Requirements:

- Develop and implement a management system, assigning a qualified person with the overall responsibility for the program
- Conduct a hazard assessment
- Implement a Program 2 Prevention Program
- Develop and implement an emergency response program
- Submit the data on prevention program elements for Program 2 processes.

Additional Program 3 Requirements:

- Develop and implement a management system, assigning a qualified person with the overall responsibility for the program
- Conduct a hazard assessment
- Implement a Program 3 Prevention Program
- Develop and implement an emergency response program
- Submit the data on prevention program elements for Program 3 processes.

40 CFR 82
Protection of Stratospheric Ozone
Subpart A: Production and Consumption Controls
Subpart E: The Labeling of Products Using
Ozone-Depleting Substances
Subpart F: Recycling and Emissions Reduction

Applicability:

Any individual, corporate or government entity that produces, transforms, imports, or exports these controlled substances.

Note: The list below is not all inclusive of Title VI requirements affecting food processors.

40 CFR 82	
Requirements	**Effective Date**
Subpart A: Production and Consumption Controls	
Prohibition on the production and consumption of any Class I substance in annual quantities greater than the relevant percentage specified in the regulations (based on quantity of substance produced in the baseline year).	January 1 of each year specified in the regulations.
Prohibition on the production of all Class I substances.	1996 (except for methyl bromide - 2001)
Prohibition on the production of all Class II substances.	Begin in 2003 and end 2030
Reporting Requirements: Reports on production, imports, and exports of Class I and II substances.	Quarterly
Subpart E: The Labeling of Products Using Ozone-Depleting Substances	
Containers in which Class I and II refrigerants are stored or transported are required to be labeled with a warning stating that it contains a substance which harms public health and environment by destroying ozone in the upper atmosphere.	
Subpart F: Recycling and Emissions Reduction	
Prohibition on knowingly venting ozone-depleting compounds used as refrigerants into the atmosphere during maintenance, service, repair, or disposal or air-conditioning or refrigeration equipment.	July 1, 1992

40 CFR 82	
Technicians servicing air-conditioning and refrigeration equipment are required to evacuate refrigerant in the line according to prescribed guidelines.	July 13, 1993
Recovery and/or recycling equipment must be tested by an EPA-approved third-party testing organization.	All equipment sold after November 15, 1993. Some equipment manufactured prior to this date is grandfathered.
Require repair of substantial leaks in all air conditioning and refrigeration units with more than 50 pounds of refrigerant.	Within 30 days of recovery
All persons who maintain, service, repair, or dispose of appliances are required to be certified.	November 14, 1994
Persons servicing or disposing of air-conditioning and refrigeration equipment are required to certify that certified recovery and recycling equipment has been acquired and they are complying with the applicable requirements of 40 CFR 82, Subpart F.	August 12, 1993

APPENDIX A.4
SUMMARY OF PRINCIPAL REGULATIONS UNDER THE EMERGENCY PLANNING AND COMMUNITY RIGHT-TO-KNOW ACT (EPCRA)

The following sections provide a summary of the principal regulations developed pursuant to EPCRA that may apply to the food processing industry. The regulations included are:

- **40 CFR 355** - Emergency Planning and Notification

- **40 CFR 370** - Hazardous Chemical Reporting: Community Right-to-Know

- **40 CFR 372** - Toxic Chemical Release Reporting, Community Right-to-Know

40 CFR 355
Emergency Planning and Notification

Emergency Planning, 40 CFR 355.30

Requirements	
Facilities subject to emergency planning requirements must notify the local emergency planning committee and State emergency response commission. They must designate an emergency planning coordinator and provide information to the local emergency planning committee.	The facility has onsite an extremely hazardous substance equal to or greater than its threshold planning quantity. (The list of extremely hazardous substances and threshold planning quantities (TPQs) is in 40 CFR 355, Appendices A and B. Section 355.30(e) tells you how to calculate TPQs for solids and mixtures.)
• 40 CFR 355.30(b) notification that the facility is subject to the planning requirements is due May 17, 1987 or within 60 days of becoming subject to the planning requirements	
• 40 CFR 355.30(c) facility emergency coordinator designated due September 17, 1987 or 30 days after a local emergency planning committee is established	
• 40 CFR 355.30(d) information for planning must be provided "promptly" upon request; notice of any relevant changes must also be provided.	

Emergency Release Notification, 40 CFR 355.40

Requirements	Regulatory Threshold
A facility must immediately notify the local community emergency coordinator (or emergency response personnel) of any area likely to be affected, and State emergency response commission of any State likely to be affected by a release. Notice must include chemical name or identity of any substance released, indication of whether it is an extremely hazardous substance, estimate of quantity released, estimate of time and duration of release, media into which release occurred, known or expected acute or chronic health risks including medical advice for exposed individuals, precautions to be taken, contact/phone numbers for further information. For transportation-related releases, this information can be provided to the 911 operator.

A written follow up emergency notice must be provided to update the information about the release, and actions taken (not required for transportation-related releases). | The facility produces, uses, or stores a hazardous chemical and there is a release of a reportable quantity of any extremely hazardous substance or CERCLA hazardous substance. (Extremely hazardous substances and reportable quantities are in 40 CFR 355, Appendices A and B. CERCLA hazardous substances are in 40 CFR 302, Table 302.4.) |

40 CFR 370
Hazardous Chemical Reporting: Community Right-to-Know

General Applicability:

Any facility that is required to prepare or have available an MSDS for a hazardous chemical under OSHA (1970).

Reporting Requirements, 40 CFR 370.20

40 CFR 370 applies to any facility that has present at any time hazardous chemicals in an amount greater than or equal to 10,000 lbs or extremely hazardous substances in an amount greater than or equal to 500 pounds, or the Threshold Planning Quantity (TPQ), whichever is less. Such facilities must submit Tier I forms by March 1, 1991 and annually thereafter. If requested, they must also submit Tier II forms.

MSDS Reporting, 40 CFR 370.21

Subject facilities must submit to the local emergency planning committee (LEPC), state emergency response commission (SERC) and the local fire department (1) MSDSs for the facility for hazardous chemicals as required in 40 CFR 370.20; or (2) similar information including a list of hazardous chemicals by hazard category, the chemical or common name and components.

Reporting Upon Request, 40 CFR 370.21(d)

An MSDS must be provided within 30 days of receipt of the request by the LEPC.

Supplemental Reporting, 40 CFR 370.21(c)

An MSDS must be provided within 3 months of : (1) discovering new information on a chemical, (2) being required to have an MSDS for a chemical, or (3) the chemical being present above threshold.

Inventory Reporting, 40 CFR 370.25

The owner or operator must provide an inventory form to the emergency planning commission, the committee and the fire department with jurisdiction over the facility. It should contain Tier I information on hazardous chemicals present at the facility during the preceding calendar year above the threshold levels in 40 CFR 370.20(b). It must be submitted on or before March 1 each year. Tier II information may be submitted as an alternative per 40 CFR 370.25(b).

Submission of Tier II Information, 40 CFR 370.25(c)

Upon request by the committee, the facility must submit Tier II information.

Fire Department Inspection, 40 CFR 370.25(d)

The facility must allow the fire department to conduct inspections and must provide specific information on locations of chemicals upon request.

Mixtures, 40 CFR 370.28

The facility must report on mixtures and quantify its mixtures using procedures in 40 CFR 370.28.

Public Access and Availability of Information (Subpart C), 40 CFR 370.30

The committee must provide any person with MSDS or Tier II information for a specific facility, except that upon request by the facility owner or operator, the commission or committee can withhold information on the locations of chemicals identified on Tier II forms.

Tier I and Tier II Inventory Forms (Subpart D), 40 CFR 370.40

Subpart D contains the Tier I and Tier II forms that are used to report information on hazardous and extremely hazardous chemicals at the facility. (Some states have their own forms.)

40 CFR 372
Toxic Chemical Release Reporting, Community Right-to-Know

General Provisions - Subpart A
Recordkeeping, 40 CFR 372.10

Applicability	
Facilities must retain copies of reports, supporting documentation, data to show how reportable quantities were determined, data to calculate the quantity of a release, documentation of offsite transfer or release of toxic chemicals, manifests or records for offsite transfer for a period of 3 years after each report is made. The reports must be available for inspection by EPA. Date of applicability: January 1, 1987 Threshold in 40 CFR 372.25(a) applies to chemicals manufacturered, imported or processed at a facility. Since the 1989, the threshold is 25,000 lb/yr. Note: thresholds apply to individual chemicals over threshold levels, or to the combined totals of more than one chemical if the combined amount exceeds a threshold. Details are presented in 40 CFR 372.25(b)-(h).	All facilities where releases have been reported or where chemicals are manufactured, imported or processed at or above TPQs.

Reporting Requirements, Subpart B

Requirements	Affected Facility
This section of the regulations sets forth requirements for the submission of information relating to the release of toxic chemicals under Section 313 of EPCRA yearly on July 1. Date of applicability: January 16, 1988.	40 CFR 372.22 specifies types of facilities that are subject to the Form R reporting requirements: a) facilities with more than 10 full-time employees, b) facilities in SIC codes 20-39 (as of January 1, 1987). Criteria for the determination of SIC codes are further explained in 40 CFR 372.22(b), and c) facilities that process, manufacture, or use a toxic chemical in excess of the threshold quantity set forth for the chemical in 40 CFR 372.30. Exemptions to the reporting of releases of toxic chemicals are detailed in 40 CFR 372.28 (e.g., *de minimis* concentrations, toxic chemicals contained inarticles, structural components, nonroutine janitorial uses, personal use by employees or for motor vehicles, chemicals in process water or noncontact cooling water, and laboratory activities). Owners of industrial parks or similar real estate owners are also exempt since the operators of the facilities would hold this responsibility.

Reporting Requirements and Schedule for Reporting, 40 CFR 372.30

Applicability	
EPA Form 9350-1 is to be used to report chemicals above thresholds for manufactured, imported, processed, used or combined into a mixture or trade name product. Details on characterizing mixtures and trade name products are given in 40 CFR 372.30(b). 40 CFR 372.30(d) Reports are due annually on July 1 beginning in 1987. Additional specific data requirements for the years 1987, 1988, and 1989 are procided in 40 CFR 372.30(e). Chemicals above threshold planning quantities.	A regulated facility may consist of more than establishment (defined as economic unit) and separate forms may be used for each establishment as long as reporting is accomplished for the entire facility.

Supplier Notification Requirement - Subpart C

Applicability	
Facilities must notify the person to whom toxic chemicals, mixtures or trade name products containing toxic chemicals are sold. The notification must be in writing and include specific information per 40 CFR 372.45(b) (product trade name, a statement thtat the product contains a SARA Title III constituent, the CASE number of the chemical, and the % by weight of the toxic chemical). Notification must be with the first shipment of the product in each calendar year. If the product is renamed or changed, the notification must be initiated over again.	Owners and operators of facilities classified as SIC code 20-39 who manufacture, import or process toxic chemicals, and who sell or otherwise distribute a mixture or trade name product containing a toxic chemical to a facility who uses of sells the product or mixture. If an MSDS is required in accordance with 29 CFR 1919.1200, the notification must be attached or incorporated into the MSDS. Exceptions include mixtures or trade name chemicals with *de minimis* amounts (see 40 CFR 372.45(d) for others). However, if the chemical is considered proprietary (trade secret) under 29 CFR 1910.1200, the notification can be written with only generic language.

Specific Toxic Chemical Listings - Subpart D
Chemicals and Chemical Categories to which 40 CFR 372.65 Applies

Applicability	
A table with alphabetical listing of categories and chemicals, including CAS numbers and effective date for each chemical is provided. Date of applicability: January 1, 1987	All facilities must characterize their chemicals.

Forms and Instructions - Subpart E
Toxic Chemical Release Reporting Form and Instruction - 40 CFR 372.85

Applicability	
See Reporting Requirements - Subpart B.	See Reporting Requirements and Schedule for Reporting, 40 CFR 372.30.

APPENDIX A.5
SUMMARY OF PRINCIPAL REGULATIONS UNDER THE COMPREHENSIVE ENVIRONMENTAL RESPONSE, LIABILITY, AND COMPENSATION ACT (CERCLA)

The following sections provide a summary of the principal regulations developed pursuant to CERCLA that may apply to the food processing industry. The regulations included are:

- **40 CFR 302** - Designation, Reportable Quantities, and Notification

40 CFR 302
Designation, Reportable Quantities, and Notification

Designation of Hazardous Substances, 40 CFR 302.4

Requirements	
Under Section 102(a) of CERCLA, these regulations identify reportable quantities of hazardous substances and set forth reporting requirements of releases. Listed hazardous substances are in 40 CFR 302, Table 302.4 and are designated as "hazardous under Section 102 (a) of CERCLA." Also included are "unlisted" hazardous substances which are defined in 40 CFR 302.4(b) as characteristics of hazardous waste.	The Table includes the reportable quantities of these substances. Unlisted hazardous substances have reportable quantity limit of 100 pounds (40 CFR 302.5), except for unlisted hazardous wastes that exhibit extraction procedure (EP) toxicity as identified in Part 261 which vary based on the reportable quantity of the pollutant of concern and its lowest value in Table 40 CFR 302.4. Appendix A of 40 CFR 302.4 contains a sequential Chemical Abstract Service (CAS) number listing of chemicals and Appendix B contains a listing of regulated radionuclides.

Notification Requirements, 40 CFR 302.6

Requirements	
Facilities which release reportable quantities established in 40 CFR 302, Table 302.4 must immediately notify the National Response Center at (800) 424-8802 or in the Washington D.C. area at (202) 426-2675. 40 CFR 302, Table 302.4 is used to determine whether the regulations apply to a specific facility based on chemicals that are released.	Exposure to persons within a workplace is excluded. Reportable quantities range from 1 to 5,000 pounds. Release means any spill, leak, pumping, emitting, emptying, discharging, injecting, escaping, leaching, dumping, or disposing into the environment. Specific requirements for various types of radionuclides, including those which are exempt from reporting to the National Response Center are given in 40 CFR 302.6.

APPENDIX A.6
SUMMARY OF PRINCIPAL REGULATIONS
UNDER THE RESOURCE CONSERVATION AND
RECOVERY ACT

The following sections provide summaries of the principal regulations developed pursuant to RCRA that may apply to the food processing industry. The section includes:

- •• **40 CFR 261.5 and 262.34** - Generator Classifications and Requirements
- •• **40 CFR 262** - Hazardous Waste Generator Requirements
- •• **40 CFR 263** - Hazardous Waste Transporter Requirements
- •• **40 CFR 268** - Land Disposal Restrictions
- •• **40 CFR 280** - Underground Storage Tanks (UST)

40 CFR 261.5 and 262.34
Generator Classifications and Requirements

Conditionally Exempt Small Quantity Generator (CESQG)

Requirements	Affected Facility
• Make hazardous waste determination under 40 CFR 262.11	• Generate 100 kg/month (220 lbs/month) or less of hazardous waste, or
• Waste must be managed and disposed in a hazardous waste facility, or a landfill or other facility approved by the State for industrial or municipal wastes	• Generate 1 kg/month (2.2 lbs/month) or less of acute hazardous waste, or
• Must comply with 40 CFR 261.5(g) to be excluded from requirements under 40 CFR 262 through 266, 268, and 270.	• Accumulate up to 1,000 kg (2,200 lbs) of hazardous waste onsite at any time

40 CFR 262
Hazardous Waste Generator Requirements

Small Quantity Generator (SQG)

Requirements	Affected Facility
• Subject to regulation under 40 CFR 262 through 266, 268, and 270. • Special requirements under 40 CFR 265.201 for accumulating hazardous waste in tanks. • May not accumulate more than 6,000 kg of hazardous waste at any time. • May not accumulate hazardous waste onsite for longer than 180 days (270 days if waste must be transported over 200 miles to hazardous waste facility), otherwise hazardous waste storage permit required.	• Generate more than 100 kg/month (220 lbs/month) of hazardous waste, but less than 1,000 kg/month (2,200 lbs/month) of hazardous waste, or • Accumulate more than 1,000 kg (2,200 lbs), but less than 6,000 kg of hazardous waste at any time

Large Quantity Generator (LQG)

Requirements	Affected Facility
• Subject to regulation under 40 CFR 262 through 266, 268, and 270. • May not store hazardous waste onsite for more than 90 days, otherwise hazardous waste storage permit required.	• Generate 1,000 kg/month (2,200 lbs/month) or more of hazardous waste, or • Generate 1 kg/month (2.2 lbs/month) or more of acutely hazardous waste, or • Generate 100 kg/month (220 lbs/month) or more of spill cleanup debris containing an acutely hazardous waste, or • Accumulate 1 kg (2.2 lbs) or more of acutely hazardous waste at any time

40 CFR 262 - HAZARDOUS WASTE GENERATOR REQUIREMENTS		
Requirements	**Description**	**Affected Facility**
EPA ID Number 40 CFR 262.12	• Cannot treat, store dispose of, or transport hazardous waste without EPA ID Number	LQG or SQG that transports, or offers for transportation, hazardous waste for offsite treatment, storage or disposal
	• Cannot offer hazardous waste to transporter or to treatment, storage, or disposal facilities that do not have an EPA ID Number	
Subpart B - Manifest Requirements 40 CFR 262.20-260.33	•• Must complete and sign EPA form 8700-22 or 8700-22A for each shipment of hazardous waste	
Subpart C - Pre-transport Requirements 40 CFR 262.30-262.34	• Must label and package hazardous waste in accordance with DOT regulations (49 CFR Parts 172, 173, 178, 179) prior to transport	
	• Accumulation in units that comply with Subpart I of 40 CFR 265 (containers), or Subpart J of 40 CFR 265 (tanks)	SQGs allowed up to 180 (or 270) days for accumulating hazardous waste without a storage permit
	• Accumulation in units that comply with air emission standards identified in 40 CFR 265 Subparts AA (process vents), BB (equipment leaks) and CC (tanks, surface impoundments and containers) and with Subpart DD (containment buildings)	May accumulate wastes up to 90 days without storage permit
	• Must develop and maintain a contingency plan for storing wastes onsite	
Subpart D - Recordkeeping and Reporting 40 CFR 262.40-262.44	• Maintain copies of manifest for three years	SQG exempt from biennial reporting requirements
	• Must prepare and submit Biennial Report	
	• Must file exception report if manifests not received by designated facility within 35 days (LQG) or 60 days (SQG)	

40 CFR 262 - HAZARDOUS WASTE GENERATOR REQUIREMENTS		
Requirements	**Description**	**Affected Facility**
Subpart E - Exports of Hazardous Waste 40 CFR 262.50-262.57 Subpart F - Imports of Hazardous Waste 40 CFR 262.60	• Notify EPA 60 days before shipment • Must confirm waste receipts or file an exception report • Must file a Summary Report of Foreign Activity on March 1 of each year • Must prepare manifest that identifies foreign generator and importer •• Must comply with all other generator standards in 40 CFR 262	

40 CFR 263
Hazardous Waste Transporter Requirements

40 CFR 263 - HAZARDOUS WASTE TRANSPORTER REQUIREMENTS		
Requirements	**Description**	
EPA ID Number 40 CFR 263.11	• Must obtain an EPA ID Number in order to transport hazardous waste	Persons who transport hazardous waste within the U.S. if manifest is required under 40 CFR Part 262.
Transfer Facility Requirements 40 CFR 263.12	• May store manifested shipments for ten days or less, otherwise subject to hazardous waste storage requirements under 40 CFR 264, 265, 268, and 270	
Manifest and Recordkeeping Requirements 40 CFR 263.20	• Cannot receive a waste shipment unless accompanied by a hazardous waste manifest	
Hazardous Waste Discharges 40 CFR 263.30	• Take appropriate action • Notify proper authorities	

40 CFR 268
Land Disposal Restrictions - Certification and Notification

40 CFR 268 - GENERATOR - CERTIFICATION AND NOTIFICATION		
Requirements	**Description[1]**	**Affected Facility**
Waste Analysis and Recordkeeping for Generators 40 CFR 268.7(a)	•• Must determine if waste is restricted from land disposal •• If waste does not meet treatment standards in 40 CFR 268 Subpart D, must provide a one-time notification to treatment or storage facility receiving waste •• If waste meets treatment standards in 40 CFR 268 Subpart D, must submit a one-time notification and certification to treatment, storage, or disposal facility receiving the waste •• If accumulating and treating restricted wastes onsite, must develop waste analysis plan and keep in files onsite •• Maintain copies of records, certifications, and notices for three years. Records may be maintained electronically	LQGs and SQGs

[1] EPA recently amended the LDR regulations. For more information, see the Federal Register Vol. 62 No. 91; May 12, 1997.

40 CFR 280 - UNDERGROUND STORAGE TANK REQUIREMENTS		
Requirements	**Description**	**Affected Facility**
Design, Construction, Installation, and Notification (Subpart B)	•• New USTs (installed after December 1988) must meet performance standards detailed in 40 CFR 280.20 •• All existing UST systems (installed before December 1988) must be upgraded to meet standards detailed in 40 CFR 280.21 by December 1998 •• Notify State and/or local agencies upon the Installation and use of new UST systems (40 CFR 280.22)	All owners and operators of underground storage tank systems as defined in 40 CFR 280.12 (See Section 280.10 (b-d) for exceptions)

40 CFR 280 - UNDERGROUND STORAGE TANK REQUIREMENTS		
Requirements	**Description**	**Affected Facility**
General Operating Requirements (Subpart C)	•• Must ensure the prevention of releases through spill and overfill control, proper corrosion protection, use of compatible materials, and proper and appropriate repairs to the UST system •• Reporting requirements include notification, reports of all releases (suspected and confirmed), corrective action, and permanent changes in service or closure. •• Recordkeeping requirements include documentation of corrosion controls, UST system repairs, release detection compliance	
Release Detection (Subpart D)	•• Must provide a method or combination of methods to detect leaks and releases from the UST system •• Must comply with release detection requirements according to the schedule set forth in 40 CFR 280.40(c) •• Petroleum USTs must comply with release detection requirements under 40 CFR 280.41 •• Hazardous substance USTs must comply with release detection requirements under 40 CFR 280.42 •• Must maintain records demonstrating compliance with release detection requirements	
Release Reporting, Investigation, and Confirmation (Subpart E)	•• Must report any suspected releases within 24 hours or another reasonable time period specified by implementing agency •• Must investigate and confirm any suspected releases •• Must contain and cleanup any release, and report to implementing agency	
Release Response and Corrective Action for UST Systems Containing Petroleum or Hazardous Substances (Subpart F)	In the event of a release: •• Must notify implementing agency upon confirmation of a release, and take action as necessary	UST systems that manage petroleum or hazardous substances.

40 CFR 280 - UNDERGROUND STORAGE TANK REQUIREMENTS		
Requirements	**Description**	
Out-of-Service UST Systems and Closure (Subpart G)	•• Must notify within 30 days of permanent closure •• Must maintain records to demonstrate compliance with closure requirements in accordance with 40 CFR 280.34	
Financial Responsibility (Subpart H)	•• Must demonstrate financial responsibility for taking corrective action and for compensating third parties for bodily injury and property damage caused by accidental releases	

APPENDIX A.7
PROPOSED AND PENDING REGULATIONS

CWA	
Pending SPCC Proposals	October 22, 1991 (56 FR 45612) February 17, 1993 (58 FR 8824) December 2, 1997 (62 FR 63811)
CAA: MACTs Affecting SIC Code 20 Facilities:	
MACTs affecting the SIC Code 20 facilities include: Aerosol Can-Filling Facilities, Baker's Yeast Manufacturing, Cellulose Food Casing Manufacturing, and Vegetable Oil Production	November 15, 2000
CEPPO's the use of NAICS codes: *Accidental Release Prevention Requirements: Rick Management Programs Under Clean Air Act Section 112 (r)(7) Amendments.*	Proposed rule: 63 FR 19216, April 17, 1998. Final rule anticipated October 1998.
SDWA	
Interim Enhanced Surface Water Treatment Rule and Stage 1 Disinfection Byproduct Rule	November 1998 - Propose Rule November 1999 - Final Rule November 2001 - Implementation
State II Disinfection Byproducts Rule	May 2002 - Propose Rule
Filter Backwash Recycling Rule	May 2002 - Final Rule
Groundwater Disinfection Rule: Issue regulations requiring disinfection for all public water systems, including surface water systems and "as necessary" groundwater systems, and promulgate criteria for determining whether to require in groundwater systems.	After August 1999 By May 2002
Long Term 1 Enhanced Surface Water Treatment Rule (LTESWTR)	November 2000 - Final Rule

APPENDIX B. RESOURCES

Introduction

This section presents lists of federal and state contacts that you can call or access via the Internet if you are seeking additional information on a particular issue or regulation. A quick reference list of EPA hotlines and Internet addresses is presented below, followed by contacts for the following:

- •• Clean Water Act (CWA)
- •• Safe Drinking Water Act (SDWA)
- •• Clean Air Act (CAA)
- •• Emergency Planning and Community Right-To-Know-Act (EPCRA)
- •• Resource Conservation and Recovery Act (RCRA)
- •• Toxic Substances Control Act (TSCA)
- •• Business policies and contacts
- •• EPA, state and local references and contacts.

Quick Reference List

Hotlines		
National Response Center Hotline		(800) 424-8802
CWA	Storm Water Hotline (NPDES Program)	(800) 245-6510
	National Small Flows Clearinghouse	(800) 624-8301
	Wetlands Information Hotline	(800) 832-7828
	National Small Flows Clearinghouse	(800) 624-8301
	Water Quality Information Center	(301) 504-6077
SDWA	Safe Drinking Water Hotline	(800) 426-4791
	National Drinking Water Clearinghouse	(800) 624-8301
CAA	Stratospheric Ozone Information Hotline	(800) 296-1996
	Air RISC Information Support Center Hotline	(919) 541-0888
	Asbestos Ombudsman Clearinghouse/Hotline	(800) 368-5888
	Control Technology Center	(919) 541-0800
	Indoor Air Quality Information Clearinghouse	(800) 438-4318 or (202) 484-1307

Hotlines		
	RCRA/UST, Superfund, and EPCRA Hotline (Provides information on the Risk Management Program under the CAA.)	(800) 424-9346 or (703) 412-9810
EPCRA	RCRA/UST, Superfund and EPCRA Hotline [Also CAA Section 112(r) - RMP]	(800) 424-9346 or (703) 412-9810
RCRA	RCRA/UST, Superfund and EPCRA Hotline [Also CAA Section 112(r) - RMP]	(800) 424-9346 or (703) 412-9810
	Hazardous Waste Ombudsman	(800) 262-7937 or (202) 260-9361
TSCA	TSCA Assistance Information Service	(202) 554-1404
Small Business	EPA's Small Business Ombudsman Clearinghouse/Hotline	(800) 368-5888 or (703) 305-6462

EPA Internet Sites

••   EPA's Home Page   http://www.epa.gov

••   EPA Public Information Center   http://www.epa.gov/PIC.html

•   EPA's Pollution Prevention Information Clearinghouse   http://www.epa.gov/opptintr/library/libppic.htm

••   Pollution Prevention Home Page   http://www.epa.gov/opptintr/p2home

••   EPA's Enviroene   http://www.epa.gov/envirosense

••   Emergency Response   http://www.epa.gov/ERNS

••   Toxic Release Inventory   http://www.epa.gov/opptintr/tri

••   ECPRA laws   http://www.epa.gov/swercepp/rules.html

••   Electronic Tools   http://www.epa.gov/swercepp/tools.html

••   EPA's Oil Program Home Page   http://www.epa.gov/oilspill

EPA Information Centers

- **EPA Main Library**
  401 M Street, SW., Room 2904
  Washington, DC 20460
  Phone: (202) 260-5921
  Fax: (202) 260-6257
  E-mail: Library-HQ@epamail.epa.gov

  The EPA's Main Library maintains environmental reference materials for EPA staff and the general public, including books, journals, abstracts, newsletters, and audio-visual materials generated by government agencies and the private sector. This library also provides access to online computer service bulletin boards and CD-ROM systems.

- **EPA Public Information Center**
  401 M Street, SW Mail Code 3404
  Washington, DC 20460
  (202) 260-5922
  Fax: (202) 260-6251
  E-mail: public-access@epamail.epa.gov
  Internet Address: http://www.epa.gov/PIC.html

  The Public Information Center (PIC) is a primary point of contact between EPA and the public. PIC refers calls and letters to the appropriate sources for technical information, and distributes a variety of general-interest items. The PIC is also a visitor center featuring environmental videos, photographic displays, CD-ROMs, and databases. Interested groups are encouraged to tour the facility as an introduction to the Agency as well as to learn about the environment. PIC's audience includes the general public; international visitors; students and educators; EPA and other government staff; and business, civic, and environmental groups.

- **EPA's Pollution Prevention Information Clearinghouse**
  401 M Street, SW Mail Code 7407
  Washington, DC 20460
  Phone: (202) 260-1023
  Fax: (202) 260-4659
  E-mail: ppic@epamail.epa.gov
  Internet address: http://www.epa.gov/opptintr/library/libppic.htm

  The Pollution Prevention Information Clearinghouse (PPIC) is a free, nonregulatory service of the U.S. EPA. PPIC is dedicated to reducing or eliminating industrial pollutants through technology transfer, education, and public awareness. A Reference and Referral Telephone Service is available to answer questions, take orders for documents distributed by PPIC, or refer callers to appropriate contacts. The Clearinghouse distributes selected EPA documents, pamphlets, information packets, and fact sheets on pollution prevention free of charge.

- **EPA's Enviroene**
  http://www.epa.gov/envirosense

  EPA's Enviroene, part of EPA's website, provides a single repository for pollution prevention, compliance assurance, and enforcement information, and databases from web sites inside and outside the EPA. Enviroene houses the web site for EPA's Office Enforcement and Compliance Assurance (OECA). It offers extensive case study collections, research funding information, abstracts of research projects conducted through the National Center for Environmental Research and Quality Assurance (NCERQA), industry and process-specific fact sheets, economic and cost/benefit information, audit and guidance materials, materials exchange, and announcements from the National Enforcement Training Institute (NETI).

Clean Water Act

- **EPA's Water Resource Center**
  401 M Street, SW   Mail Code: RC4100
  Washington, DC 20460
  Telephone: (202) 260-7786
  Fax: (202) 260-0386
  E-mail: waterpubs@epamail.epa.gov

  The Water Resource Center (WRC) distributes documents and other materials produced by the Office of Ground Water and Drinking Water, the Office of Science and Technology, and the Office of Wastewater Management, and provides limited research assistance to locate materials from these offices.

- **EPA's Oil Spill Program**
  To access the EPA's Oil Spill Program Information Line, call the RCRA/UST, Superfund and EPCRA Hotline at:
     (800) 424-9346 or (703) 412-9810
     (800)-535-7672 TDD line for the hearing-impaired
     (703) 412-3323 TDD in DC area
  Fax: (703) 603-9234
  To report an oil or hazardous substance release, call the National Response Center at (800) 424-8802.
  Internet address: http://www.epa.gov/oilspill

  EPA's Oil Spill Program is designed to prevent oil spills, as well as prepare for and respond to any oil spill affecting the inland waters of the U.S. The program is administered by EPA Headquarters and the 10 EPA Regions. For more information, access the web address listed above.

- ***Storm Water Hotline (Region 6 Multi-Sector General Permit (MSGP) Hotline)***
  Telephone: (800) 245-6510

  The Region 6 Multi-Sector General Permit Hotline serves as a clearinghouse for information concerning U.S. EPA Storm Water General Permits. Information specialists are available to answer technical questions concerning permit eligibility, specific permit requirements, and more. In addition, callers may order many storm water-related documents and guidance manuals.

- ***Wetlands Information Hotline***
  Telephone: (800) 832-7828

  This hotline, working in connection with EPA's Wetlands Preservation Division, responds to requests for information regarding the values and functions of wetlands and options for their protection.

- ***National Small Flows Clearinghouse***
  West Virginia University
  P.O. Box 6064
  Morgantown, WV 26506
  Telephone: (800) 624-8301
  This hotline provides information and technical assistance to help small communities reach practical, affordable solutions with their wastewater treatment problems.

- ***Water Quality Information Center***
  Telephone: (301) 504-6077

  A government clearinghouse for information on water quality and agriculture.

- **Water Environment Federation**
  601 Wythe Street
  Alexandria, VA 22314-1994
  Telephone: (800) 666-0206 or (703) 684-2452
  Fax: (703) 684-2492

  The Water Environment Federation (WEF) is an international not-for-profit educational and technical organization of over 40,000 water experts. WEF has guided technological developments in water quality and provides its members and the public with the latest information on wastewater treatment and water quality protection.

Safe Drinking Water Act

•• **Safe Drinking Water Hotline**
401 M Street, SW Mail Code 4604
Washington, DC 20460
Telephone: (800) 426-4791
Fax: (703) 285-1101
E-mail: hotline-sdwa@epamail.epa.gov

This SDWA hotline provides information about EPA's drinking water regulations and other related drinking water and ground water topics to the regulated community, State and local officials, and the public. Specifically, the Hotline clarifies drinking water regulations, provides appropriate 40 CFR and Federal Register citations, explains EPA-provided policies and guidelines and gives update information on the status of regulations. The Hotline can also provide State and local contacts. The Hotline can take orders for EPA drinking water publications or (if the publication is not available from the Office of Water) refer callers to the appropriate ordering organization.

• *National Drinking Water Clearinghouse*
West Virginia University
P.O. Box 6064
Morgantown, WV 26506
Telephone: (800) 624-8301

This clearinghouse provides free and low-cost technical assistance and information on small community drinking water systems.

•• *The American Water Works Association*
1401 New York Ave., N.W., Suite 640
Washington, DC 20005
Telephone: (202) 628-8303

The American Water Works Association (AWWA) is an international nonprofit scientific and educational society dedicated to the improvement of drinking water quality and supply. Its more than 50,000 members represent the full spectrum of the drinking water community: treatment plant operators and managers, scientists, environmentalists, manufacturers, academicians, regulators, and others who hold genuine interest in water supply and public health. Membership includes more than 3,700 utilities that supply water to roughly 170 million people in North America.

WaterWiser is the AWWA's clearinghouse on water efficiency information.
6666 West Quincy Avenue
Denver, CO 80235
Telephone: (800) 559-9855

Clean Air Act

- ***Stratospheric Ozone Information Hotline***
  Telephone: (800) 296-1996
  Fax: (202) 775-6681

  The Stratospheric Ozone Information Hotline provides in-depth information on ozone protection regulations and requirements under Title VI of the Clean Air Act Amendments of 1990. Information on the transition to non-ozone-depleting chemicals in various use sectors, as well as retrofitting equipment and refrigerant management is also available. In addition, the Hotline serves as a distribution center and point of referral for an array of information pertaining to other general aspects of stratospheric ozone depletion and its protection. The Hotline maintains a library of relevant policy and science documents, reports, articles, and contact lists.

- ***Asbestos Ombudsman Clearinghouse/Hotline***
  Telephone: (800) 368-5888 or (703)305-5938
  Fax: (703) 305-6462

  The Asbestos Ombudsman Clearinghouse/Hotline responds to requests for information relating to the handling and abatement of asbestos in schools, the workplace, and the home. The information is available to the public, including individual citizens and those involved in community services. In addition, the hotline offers assistance to small businesses in complying with EPA regulations.

- ***Control Technology Center***
  Telephone: (919) 541-0800

  The CTC offers technical support and guidance concerning air pollution emissions pollution prevention, and control technology for all air pollutants, including air toxins emitted by stationary sources.

- ***Indoor Air Quality Information Clearinghouse***
  P.O. Box 37133
  Washington, DC 20013-7133
  Telephone: (800) 438-4318, (202) 484-1307
  Fax: (202) 484-1510
  E-mail: iaqinfo@aol.com
  Internet address: http://www.epa.gov/iedweb00/iaqinfo.html

  The Indoor Air Quality Information Clearinghouse (IAQ INFO) provides access to public information on indoor environments through a range of services including: an operator-assisted hotline; distribution of relevant EPA publications at no charge; literature searches on a topic for further reference; referrals to appropriate government agencies, research, public interest, and industry representatives; and information about training courses and materials. IAQ INFO responds to inquiries from the general public, state and local government officials, service providers, educators, and health professionals. IAQ INFO provides information on many aspects of indoor air quality including pollutants such as carbon monoxide, radon, and biological contaminants; problem prevention approaches; testing, measurement and mitigation; and legislation, guidelines, and voluntary programs.

- **Air RISC Information Support Center Hotline**
  Office of Air Quality Planning and Standards, MD-15
  Research Triangle Park, NC 27711
  Telephone: (919) 541-0888, 541-5741
  Fax: (919) 541-0824, 541-2045
  Internet address: http://www.epa.gov/oar/oaq_ttn.html

  The Air RISC provides technical assistance and information in areas of health, risk, and exposure assessment for toxic and criteria air pollutants. Services include the hotline for direct access to EPA experts; detailed technical assistance for more in-depth evaluations or information; and general technical guidance in the form of documents, reports and training materials related to health, risk and exposure assessment. Air RISC can also be accessed through the OAQPS Technology Transfer Network (TTN).

- - **RCRA/UST, Superfund and EPCRA Hotline**

  This hotline provides information on the Risk Management Program under the Clean Air Act (CAA) Section 112(r). See telephone numbers and an additional description of this hotline in the next item below.

Emergency Planning and Community Right-To-Know Act

- **RCRA/UST, Superfund and EPCRA Hotline**
  Telephone: (800) 424-9346 or (703) 412-9810
                       (800)-535-7672 TDD line for the hearing-impaired
                       (703) 412-3323 TDD in DC area
  Fax: (703) 603-9234

  This hotline provides information about the regulations and programs implemented under the Resource Conservation and Recovery Act (RCRA), the Comprehensive Environmental Response Compensation and Liability Act (CERCLA or Superfund), the Emergency Planning, Community Right-to-Know Act (EPCRA)/ Superfund Amendments Reauthorization Act (SARA Title III), and the Risk Management Program under the Clean Air Act (CAA) Section 112(r). This hotline also provides referrals for documents related to these programs. Translation is available for Spanish-speaking callers.

Resource Conservation and Recovery Act

- **RCRA/UST, Superfund and EPCRA Hotline**
  Telephone: (800) 424-9346 or (703) 412-9810
                 (800)-53 -7672 TDD line for the hearing-impaired
                 (703) 412-3323 TDD in DC area

Fax: (703) 603-9234
E-mail: RCRA-Docket@epamail.epa.gov
Internet address: http://www.epa.gov/epaoswer/hotline/index.htm

This hotline answers factual questions about EPA regulations and programs under RCRA, Superfund, and EPCRA, and responds to requests for relevant documents. Specifically, the Hotline responds to inquiries about waste minimization programs required under RCRA, source reduction and hazardous waste combustion, Section 6607 of the Pollution Prevention Act of 1990, which expanded data collection under EPCRA 313, and other components of the waste management regulatory programs. This hotline also provides referrals for documents related to these programs. Translation is available for Spanish-speaking callers.

- **Hazardous Waste Ombudsman**
  Telephone: (800) 262-7937 or (202) 260-9361

  The Hazardous Waste Management Program, established under the Resource Conservation and Recovery Act (RCRA) assists the public and the regulatory community in resolving problems concerning any program or requirement under the Hazardous Waste Program. The ombudsman handles complaints from citizens and the regulatory community, conducts investigations, undertakes site reviews, issues reports, and serves as a national resource for the newly appointed ombudsman in the 10 EPA Regions. The ombudsman also provide information on the Risk Management Program under the Clean Air Act (CAA) Section 112 (r).

Toxic Substances Control Act

- **Toxic Substances Control Act (TSCA) Assistance**
  Telephone: (202) 554-1404
  Fax: (202) 554-5603
  Email: tscahotline@epamail.epa.gov

  The information service furnishes TSCA regulation information to the chemical industry, labor and trade organizations, environmental groups, and the general public. Technical as well as general information is available.

Business Policies and Contacts

Business Policies

•• **Small Business Regulatory Enforcement Fairness Act (SBREFA) of 1996**

This Act (Title II of Public Law 104-121) was signed into law by President Clinton on March 29, 1996 in order to accomplish several key objectives: (1) to encourage small businesses to participate more in the Federal regulatory process; (2) to require Federal agencies to become more responsive to small business inquiries relating to regulatory and reporting requirements; (3) to promote greater cooperation between small businesses and government agencies through less punitive and more solution-oriented approaches to environmental compliance; and (4) to make federal agencies more accountable for excessive enforcement actions by permitting small businesses to go to court to be awarded attorney's fees and costs.

For more information about the Small Business Regulatory Enforcement Fairness Act (SBREFA), access the following Internet address: http://www.sba.gov/regfair/overview.html/.

•• *Final Policy on Compliance Incentives for Small Businesses*

This policy implements Section 323 of the Small Business Regulatory Enforcement Fairness Act (SBREFA) of 1996. This Policy, effective June 10, 1996, is intended to promote environmental compliance among small businesses by providing them with special incentives to participate in compliance assistance programs or to conduct environmental audits, and to help them promptly correct violations. The federal or state governments can waive all or a portion of a penalty when a violation is (1) identified through government-supported on-site compliance assistance or through an environmental audit, (2) disclosed, and (3) corrected. Under this policy, a "small" business is defined as having 100 or fewer employees across all facilities and operations owned by the entity. There is no limitation on the amount of pollutants produced by the small business.

You can obtain more information on this policy and how it may apply to your business by reviewing the final policy in the Federal Register [61 FR 27984 (June 3, 1996)] or by accessing EPA's Enviroene site at: http://es.inel.gov/oeca/smbusi.html/.

• *Incentives for Self Policing: Discovery, Disclosure, Correction and Prevention of Violations Policy*

This policy mitigates or eliminates penalties for all entities, including small and large businesses, who voluntarily discover, disclose and correct violations of environmental regulations by conducting environmental audits or implementing systematic procedures reflecting the facility's due diligence.

You can obtain additional information on this self-policing policy by reviewing the final policy in the Federal Register [60 FR 66706-66711(December 22, 1995)] or by accessing EPA's Environene site at: http://es.inel.gov/oeca/ore/aed/comp/acomp/a26.html/.

•• *The 507 Small Business Enforcement Policy*

Section 507 of the 1990 Clean Air Amendments required the States to establish Small Business Stationary Source Technical and Environmental Compliance Assistance Programs. Because of State concerns that small businesses would not ask for compliance assistance if violations resulted in enforcement actions, the EPA issued the "Enforcement Response Policy for Treatment of Information Obtained Through Clean Air Act Section 507 Small Business Assistance Programs (August 12, 1994)." This policy was designed to encourage small businesses to seek compliance assistance from the States by offering two types of incentives. The first incentive allows small businesses that receive compliance assistance up to 90 days, with the possibility of an additional 90 day extension, to correct any violations discovered under the Small Business Assistance Program (SBAP). The second incentive offers compliance assistance on a confidential basis, as long as: a) the State can investigate and/or take enforcement action independent of the Section 507 program if violations are discovered; and b) confidential assistance can only be offered through SBAPs that operate independently of the state's regulatory enforcement program.

You can obtain additional information on Section 507 by accessing EPA's Environene sites at: http://es.inel.gov/oeca/ccsmd/file3.html or http://es.inel.gov/oeca/ccsmd/file11.html/.

Small Business Hotlines and Contacts

- ***EPA's Small Business Ombudsman Clearinghouse/Hotline***
  401 M Street, SW Mail Code: 1230C
  Washington, DC 20460
  Telephone: (800) 368-5888 or (703) 305-5938
  Fax: (703) 305-6462

  The mission of the EPA Small Business Ombudsman Clearinghouse/Hotline is to provide information to private citizens, small communities, small business enterprises, and trade associations representing the small business sector regarding regulatory activities. Mailings are made to update the audience on recent regulatory actions. Special attention is directed to apprising the trade associations representing small business interests with current regulatory developments. Technical questions are answered following appropriate contacts with program office staff members. Questions addressed cover all media program aspects within EPA. Inquiries are received by mail, telephone, and fax.

State Small Business Assistance Programs (SBAPs)

- For a list of state resources for small businesses and links to other government information centers (such as the Small Business Administration), access the following Internet address: http://www.epa.gov/ttn/sbap/related.html/. For more information available by phone and mail, contact the Small Business Ombudsman listed above.

- Additional information for small businesses can be found at the following EPA internet address: http://www.epa.gov/epahome/smallbus.htm/.

EPA, State and Local References and Contacts

EPA Headquarters

Environmental Protection Agency
401 M St, SW
Washington, DC 20460
Telephone: (202) 260-2090

Regional US Environmental Protection Agency Contacts

Region 1 (CT, MA, ME, NH, RI, VT)
One Congress Street
John F. Kennedy Building
Boston, MA 02203-0001
Phone: (617) 565-3420
Fax: (617) 565-3660
Toll Free: (888) 372-7341
Website: http://www.epa.gov/region01/

Region 2 (NJ, NY, PR, VI)
290 Broadway, New York
NY 10007-1866
Phone: (212) 637-3000
Fax: (212) 637-3526
Website: http://www.epa.gov/region2/

Region 3 (DC, DE, MD, PA, VA, WV)
1650 Arch Street
Philadelphia, PA 19103-2029
Phone: (215) 814-5000
Fax: (215) 814-5103
Toll free: (800) 438-2474
Website: http://www.epa.gov/region03/

Region 4 (AL, FL, GA, KY, MS, NC, SC, TN)
Environmental Protection Agency
Atlanta Federal Center
61 Forsyth Street, SW
Atlanta, GA 30303-3104
Phone: (404) 562-9900
Fax: (404) 562-8335
Toll free: (800) 241-1754
Website:http://www.epa.gov/region4/reg4.html/

Region 5 (IL, IN, MI, MN, OH, WI)
Environmental Protection Agency
77 West Jackson Boulevard
Chicago, IL 60604-3507
Phone: (312) 353-2000
Fax: (312) 353-1155
Toll free: (800) 621-8431
Website: http://www.epa.gov/region5/

Region 6 (AR, LA, NM, OK, TX)
Environmental Protection Agency
Fountain Place 12th Floor, Suite 1200
1445 Ross Avenue
Dallas, TX 75202-2733
Phone: (214) 665-2200
Fax: (214) 665-2146
Toll free: (800) 887-6063
Website:http://www.epa.gov/earth1r6/index.htm/

Region 7 (IA, KS, MO, NE)
Environmental Protection Agency
726 Minnesota Avenue
Kansas City, KS 66101
Phone: (913) 551-7003
Fax: (913) 551-7467
Toll free: (800) 223-0425
Website: http://www.epa.gov/rgytgmj/

Region 8 (CO, MT, ND, SD, UT, WY)
Environmental Protection Agency
999 18th Street Suite 500
Denver, CO 80202-2466
Phone: (303) 312-6312
Fax: (303) 312-7061
Toll free: (800) 227-8917
Website: http://www.epa.gov/unix0008/

Region 9 (AZ, CA, HI, NV)
Environmental Protection Agency
75 Hawthorne Street
San Francisco, CA 94105
Phone: (415) 744-1305
Fax: (415) 744-1070
Website: http://www.epa.gov/region09/

Region 10 (AK, ID, OR, WA)
Environmental Protection Agency
1200 6th Avenue
Seattle, WA 98101
Phone: (206) 553-1200
Fax: (206) 553-6984
Toll free: (800) 424-4372
Website: http://www.epa.gov/r10earth/

State References

•• "Sourcebook of State Laws & Regulations for Food Processors"
National Food Processors Association
1401 New York Avenue, NW
Washington, DC 20005
Phone: (202) 639-5954

This report helps food companies understand the industry's political, legal, and regulatory environment in every state. The report covers topics such as food labeling and standards, food packaging, food safety, food taxes, product liability, and school lunch programs. The report also helps the strategic planning of companies in the food industry to operate and site their plants most efficiently and help shape legislation affecting the food processing industry.

State and Local Contacts

• Information on State environmental agencies is provided below. Links to all state environmental agencies can be accessed at the Environmental Professional's Homepage at http://www.clay.net/.

• State Air Pollution Agencies: State and Territorial Air Pollution Administrators (STAPPA) and Association of Local Air Pollution Control Officials (ALAPCO). This website contains links to state government agency home pages and other state government resources and can be accessed at http://www.4cleanair.org/.

- State Emergency Response Commissions and Local Emergency Planning Committees. This website, which contains a listing of SERCs and LEPCs, can be accessed at http://www.epa.gov/swercepp/state.html/. A list of SERC and LEPC contacts is provided below.

State Environmental Protection Agencies

Alabama
AL Conservation and Natural Resources Department
PO Box 301501
Montgomery, AL 36130
Phone: (334) 240-9500
Fax: (334) 240-3380
Website: http://dcnr.state.al.us/

AL Department of Environmental Management
1751 Cong. W. L. Dickinson Drive
PO Box 301463
Montgomery, AL 36109-1463
Phone: (334) 271-7700
Fax: (334) 271-7950
Website: http://www.adem.state.al.us/

Alaska
AK Department of Environmental Conservation
410 Willoughby Avenue, Suite 105
Juneau, AK 99801-1795
Phone: (907) 465-5060
Fax: (907) 465-5070
TTY: (907) 465-5040
Website: http://www.state.ak.us/local/akpages/env.conserv/home.htm/

Arizona
AZ Department of Environmental Quality
3033 N. Central Avenue
Phoenix, AZ 85012
Phone: (602) 207-4300
Toll Free: (800) 234-5677
TTY: (602) 207-4829
Website: http://www.adep.state.az.us/

Arkansas
AR Department of Pollution Control and Ecology
8001 National Drive, PO Box 8913
Little Rock, AR 72209
Phone: (501) 682-0744
Fax: (501) 682-0798
Website: http://adeq.state.ar.us/

California
CA Environmental Protection Agency
555 Capitol Mall, Suites 235 & 525
Sacramento, CA 95814
Phone: (916) 445-3846
Website: http://calepa.ca.gov/

Colorado
Department of Natural Resources
1313 Sherman Street, Room 718
Denver, CO 80203
Phone: (303) 866-3311
Fax: (303) 866-2115

CO Department of Public Health & Environment (CDPHE)
4300 Cherry Creek Drive South
Denver, CO 80222-1530
Phone: (303) 692-2000
Fax: (303) 782-0095
TTY: (303) 691-0770
Website: http://www.state.co.us/gov_dir/cdphe_dir/cdphe.hom.html/

Connecticut
CT Department of Environmental Protection
79 Elm Street
Hartford, CT 06106-5127
Phone: (860) 424-3000
Fax: (860) 424-4053
Website: http://.dep.state.ct.us/

Delaware
DE Department of Natural Resources and
Environmental Control
89 Kings Highway, PO Box 1401
Dover, DE 19903
Phone: (302) 739-5823
Fax: (302) 739-6242
Website: http://www.dnrec.state.de.us/

District of Columbia
DC Environmental Regulation
Administration
2100 Martin Luther King Avenue, SE
Suite 203
Washington, DC 20020
Phone: (202) 645-6617
Fax: (202) 645-6102

Florida
FL Department of Environmental Protection
3900 Commonwealth Boulevard
Tallahassee, FL 32399-3000
Phone: (850) 488-1073
Fax: (850) 921-6227
Website: http://www.dep.state.fl.us

Georgia
GA Department of Natural Resources
Environmental Division
205 Butler Street, SE, Suite 1152 East
Tower
Atlanta, GA 30334-1703
Phone: (404) 656-3500
Fax: (404) 651-5778
Website: http://www.ganet.org/dnr/

Hawaii
HI Land and Natural Resources Department
Kalanimoku Building
151 Punchbowl Street, PO Box 621
Honolulu, HI 96809
Phone: (808) 587-0330
Fax: (808) 587-0390
Website: http://www.state.hi.us/dlnr/

HI Department of Health
1250 Punchbowl Street, 3rd Floor
PO Box 3378
Honolulu, HI 96801
Phone: (808) 586-4400
Fax: (808) 586-4444
Website: http://www.hawaii.gov/health/

Idaho
ID Division of Environmental Quality
1410 N. Hilton
Boise, ID 83706
Phone: (208) 373-0502
Fax: (208) 373-0417

Illinois
IL Environmental Protection Agency
1021 North Grant Avenue East
PO Box 19276
Springfield, IL 62794-9276
Phone: (217) 782-3397
Fax: (217) 785-7725
Website: http://www.epa.state.il.us/

Indiana
Indiana Department of Environmental
Management
105 N. Senate, PO Box 6015
Indianapolis, IN 46206-6015
Phone: (317) 232-8560
Toll Free: (800) 451-6027
TTY: (317) 233-6087
Website: http://www.ai.org/idem/index.html/

IN Department of Natural Resources
402 West Washington Street, Rm. W256
Indianapolis, IN 46204
Phone: (317) 233-3046
Fax: (317) 232-8036
Website: http://www.ai.org/dnr/index.html/

Iowa
IA Department of Natural Resources
Wallace State Office Building
900 E. Grand Avenue
Des Moines, IA 50319-0034
Phone: (515) 281-5145
Fax: (515) 281-8895
TTY: (515) 242-5967
Website: http://www.state.ia.us/government
/dnr/index.html/

Kansas
KS Department of Health & Environment
Landon State Office Building
900 SW Jackson Street
Topeka, KS 66612-1290
Phone: (785) 296-1522
Fax: (785) 368-6368
Website: http://www.ink.org/public/kdhe/

Kentucky
KY Natural Resources and Environmental
Protection Cabinet
Capital Plaza Tower, 5th Floor
500 Mero Street
Frankfort, KY 40601
Phone: (502) 564-3350
Website: http://www.state.ky.us/agencies
/nrepc/nrhome.htm/

KY Environmental Quality Commission
14 Reilly Road
Frankfort, KY 40601
Phone: (502) 564-2150
Fax: (502) 564-4245
Website: http://www.state.ky.us/agencies
/eqc/eqc.html/

Louisiana
LA Department of Environmental Quality
7290 Bluebonnet Boulevard, PO Box 82231
Baton Rouge, LA 70810
Phone: (504) 765-0741
Fax: (504) 765-0746
Toll Free: (888) 763-5424
24-Hour Hotline: (504) 342-1234
Website: http://www.deq.state.la.us/

LA Natural Resources Department
625 N. 4th Street, PO Box 94396
Baton Rouge, LA 70804-9396
Phone: (504) 342-4500
Fax: (504) 342-2707
Website: http://www.dnr.state.la.us/

Maine
ME Department of Environmental Protection
17 State House Station
Augusta, ME 04333-0017
Phone: (207) 287-7688
Toll Free: (800) 452-1942
Fax: (207) 287-2814
Web site: http://www.state.me.us/dep/

ME Conservation Department
22 State House Station
Augusta, ME 04333
Phone: (207) 287-2211
Fax: (207) 287-2400
TTY: (207) 287-2213

Maryland
MD Department of Environment
2500 Broening Highway
Baltimore, MD 21224
Phone: (410) 631-3000
Fax: (410) 631-3936
Toll Free: (800) 633-6101
TTY: (410) 631-3009
Website: http://www.mde.state.md.us/

Massachusetts
MA Department of Environmental Protection
1 Winter Street, 2nd Floor
Boston, MA 02108
Phone: (617) 292-5500
Fax: (617) 574-6880
Website: http://www.magnet.state.ma.us
/dem/

Michigan

MI Department of Environmental Quality
333 South Capital Avenue, PO Box 30457
Lansing, MI 48909-7957
Phone: (800) 662-9278
Fax: (517) 335-4729
Pollution Emergency Alerting System: (800) 292-4706
Website: http://www.deq.state.mi.us/

MI Natural Resources Department
Stevens T. Mason Building
530 W. Allegan Street, PO Box 30028
Lansing, MI 48909
Phone: (517) 373-1214
Fax: (517) 353-1547
Website: http://www.dnr.state.mi.us/

Minnesota

MN Department of Natural Resources
500 Lafayette Road
St. Paul, MN 55155-4040
Phone: (612) 296-6157
Fax: (612) 297-3618
Toll Free: (888) 646-6367
Website: http://www.drn.state.mn.us/

MN Office of Environmental Assistance
520 Lafayette Road North, 2nd Floor
St. Paul, MN 55155-4100
Phone: (612) 296-3417
Fax: (612) 215-0246
Toll Free: (800) 657-3843
Website: http://www.moea.state.mn.us/

Mississippi

MS Dept. of Environmental Quality
2380 Highway 80 West, PO Box 20305
Jackson, MS 39289-1305
Phone: (601) 961-5171
Fax: (601) 961-5349
Website: http://deq.state.ms.us/

Missouri

MO Department of Natural Resources
205 Jefferson Street, PO Box 176
Jefferson City, MO 65101
Phone: (573) 751-4422
Fax: (573) 751-7749
Website:
http://www.dnr.state.mo.us/homednr.htm/

Montana

MT Department of Environmental Quality
1520 E. 6th Avenue, PO Box 200901
Helena, MT 59620
Phone: (406) 444-2544
Fax: (406) 444-4386
Website: http://www.deq.state.mt.us/

MT Natural Resources and Conservation
Department
1625 11th Avenue, PO Box 201601
Helena, MT 59620-1601
Phone: (406) 444-2074
Fax: (406) 444-2684
TTY: (406) 444-6873
Website: http://www.dnrc.state.mt.us/

Nebraska

NE Natural Resources Commission
301 Centennial Mall South, PO Box 94876
Lincoln, Nebraska 68509
Telephone: (402) 471-2081
Fax: (402) 471-3132
Website: http://www.nrc.state.ne.us/

NE Department of Environmental Quality
1200 N. Street, Suite 400, PO Box 98922
Lincoln, NE 68509-8922
Phone: (402) 471-2186
Fax: (402) 471-2909
Website: http://www.deq.state.ne.us/
E-mail: pubinfo@mail.deq.state.ne.us

Nevada

NV Division of Environmental Protection
333 W. Nye Lane, Suite 138
Carson City, NV 89706-0851
Phone: (702) 687-4670
Fax: (702) 687-5856
Toll Free: (800) 992-0900

New Hampshire
NH Department of Environmental Services
Six Hazen Drive
Concord, NH 03301
Phone: (603) 271-3503
Fax: (603) 271-2867
TTY: (800) 735-2964
Website:
http://www.state.nh.us/des/descover.htm/
E-mail: pip@des.state.nh.us

New Jersey
NJ Department of Environmental Protection
401 E. State Street, 7th Floor, East Wing
PO Box 402
Trenton, NJ 08625 -0402
Phone: (609) 292-2885
Fax: (609) 292-7695
Website: http://www.state.nj.us/dep/

New Mexico
NM Environment Department
Harold S. Runnels Building
1190 S. St. Francis Drive, PO Box 26110
Santa Fe, NM 87505-4182
Phone: (505) 827-2855
Fax: (505) 827-2836
Toll Free: (800) 879-3421
Website: http://www.nmenv.state.nm.us/

New York
NY Department of Environmental
Conservation
50 Wolf Road
Albany, NY 12233
Phone: (518) 457-5400
Fax: (518) 457-7735
Website: http://www.dec.state.ny.us/

North Carolina
NC Dept. of Environment, Health, & Natural
Resources
512 North Salisbury Street, PO Box 27687
Raleigh, NC 27604
Phone: (919) 733-4984
Fax: (919) 715-3060
Website: http://www.ehnr.state.nc.us/

North Dakota
ND State Water Commission
900 East Boulevard
Bismarck, ND 58505
Phone: 701-328-2750
Fax: 701-328-3696
Website: http://www.swc.state.nd.us/

ND Environmental Health Section
1200 Missouri Avenue, PO Box 5520
Bismark, ND 58506-5520
Phone: (701) 328-5150
Fax: (701) 328-5200
Website: http://www.health.state.nd.us/

Ohio
OH Environmental Protection Agency
1800 Watermark Drive, PO Box 1049
Columbus, OH 43216-1049
Phone: (614) 644-3020
Fax: (614) 644-2329
Website: http://www.epa.state.oh.us/

OH Natural Resources Department
Fountain Square
1930 Belcher Drive
Columbus, OH 43224-1387
Phone: (614) 265-6565
Fax: (614) 261-9601
Website: http://www.dnr.state.ohio.us/

Oklahoma
OK Department of Environmental Quality
707 North Robinson, PO Box 1677
Oklahoma City, OK 73101-1677
Phone: (405) 720-6100
Toll Free: (800) 869-1400
Complaints Hotline: (800) 522-0206
Website: http://www.deq.state.ok.us/

Oregon
OR Department of Environmental Quality
811 SW Sixth Avenue
Portland, OR 97204-1390
Phone: (503) 229-5696
Fax: (503) 229-6124
Toll Free: (800) 452-4011
TTY: (503) 229-6993
Website: http://www.deq.state.or.us/

Pennsylvania
PA Department of Environmental Protection
Rachel Carson State Office Bldg.
16th Floor
400 Market Street, PO Box 2063
Harrisburg, PA 17101-2063
Phone: (717) 787-1323
Fax: (717) 783-8926
TTY: (800) 654-5984
Website: http://www.dep.state.pa.us/
E-mail: depinfo@al.dep.state.pa.us

Rhode Island
RI Department of Environmental
Management
235 Promenade Street
Providence, RI 02908
Phone: (401) 222-6800
Fax: (401) 222-3810
After-Hours Enforcement Hotline: (401) 222-2284

South Carolina
SC Dept. of Health and Environmental
Control
2600 Bull Street
Columbia, SC 29201
Phone: (803) 734-5360
Fax: (803) 734-4339
Website: http://www.state.sc.us/dhec/

SC Environmental and Natural Resources
Department
Rembert C. Dennis Building, PO Box 167
Columbia, SC 29202
Phone: (803) 734-3888
Fax: (803) 734-6951
Website: http://www.dnr.state.sc.us/

South Dakota
SD Dept. of Environment & Natural
Resources
Joe Foss Building
523 E. Capital Avenue
Pierre, SD 57501-3181
Phone: (605) 773-3151
Fax: (605) 773-4068
Website: http://www.state.sd.us/state
/executive/denr/denr.html/

Tennessee
TN Department of Environment and Natural
Resources
21st Floor, L&C Tower
401 Church Street
Nashville, TN 37243
Phone: (615) 532-0109
Fax: (615) 532-0120
Toll Free: (888) 891-8332
Website: http://www.state.tn.us/environment/

Texas
TX Natural Resource & Conservation
Commission
12100 Park 35 Circle (MC 100)
PO Box 13087
Austin, TX 78711-3087
Phone: (512) 239-4000
Fax: (512) 239-4007
Toll Free: (800) 687-4040
Website: http://www.tnrcc.state.tx.us/

Utah
UT Department of Environmental Quality
168 North 1950 W, PO Box 144810-4810
Salt Lake City, UT 84114-4810
Phone: (801) 536-4400
Fax: (801) 536-4401
Toll Free: (800) 458-0145
TTD: (801) 538-4414
Website: http://www.eq.state.ut.us/
E-mail: deqinfo@deq.state.ut.us

Vermont
VT Agency of Natural Resources
103 S. Main Street, Center Building
Waterbury, VT 05671
Phone: (802) 241-3600
Fax: (802) 244-1102
TTY: (802) 253-0191
Website: http://www.anr.state.vt.us/

Virginia
VA Department of Environmental Quality
629 East Main Street, PO Box 10009
Richmond, VA 23219
Phone: (804) 698-4000
Fax: (804) 698-4500
Toll Free: (800) 592-5482
Website: http://www.deq.state.va.us/

VA Secretary of Natural Resources
733 Ninth Street Office Building
PO Box 1475
Richmond, VA 2321
Phone: (804) 786-0044
Fax: (804) 371-8333
TTY: (804) 786-7765
Website: http://dit1.state.va.us/~snr/

Washington
WA Department of Ecology
300 Desmond Drive, PO Box 47600
Olympia, WA 98504-7600
Phone: (360) 407-7004
Fax: (360) 407-6989
TTY: (360) 407-7155
Website: http://www.wa.gov/ecology/

WA Natural Resources Department
1111 Washington Street, SE
PO Box 47001
Olympia, WA 98504-7001
Phone: (360) 902-1000
Website: http://www.wa.gov/dnr/

West Virginia
WV Division of Environmental Protection
10 McJunkin Road
Nitro, WV 25143-2506
Phone: (304) 759-0515
Fax: (304) 759-0515
TTY: (800) 637-5893
Website: http://www.192.243.139.248/

Wisconsin
WI Department of Natural Resources
101 South Webster Street, PO Box 57921
Madison, WI 53707
Phone: (608) 266-2621
Fax: (608) 267-3579
Website: http://www.dnr.state.wi.us/

Wyoming
WY Department of Environmental Quality
Herschler Building, 4th Floor
122 West 25th Street
Cheyenne, WY 82002
Phone: (307) 777-7758
Fax: (307) 777-7682
Website: http://deq.state.wy.us/ **Virgin Islands**

VI Department of Planning and Natural
Resources
396-1 Annas Retreat, Foster Building
Charlotte Amalie, US VI 00802
Phone: (340) 774-3320
Fax: (340) 775-5706
Website: http://www.gov.vi/pnr/

Puerto Rico
PR Natural and Environmental Resources
Department
Avenida Munoz Rivera, Piso 2, Parada 3 ½
PO Box 5887
San Juan, PR 00906
Phone: (787) 723-1464
Fax: (787) 723-4255
Website: http://www.fortaleza.govpr.org/
/cgi-bin/detalles.idc?Index=77

State Emergency Response Commissions (SERCs)

Alabama
Lee Helms, Co-Chair
AL Emergency Response Commission
AL Emergency Management Agency
5898 Country Road 41
PO Drawer 2160
Clanton, AL 35046-2160
Phone: (205) 280-2234
Section(s): None

James War, Co-Chair
Field Operations Office
AL Emergency Response Commission
AL Department of Environmental
Management
1751 Congressman W.L. Dickinson Drive
PO Box 301463
Montgomery, AL 36130-1463
Phone: (334) 271-7710
Section(s): None

Edward Poolos
AL Emergency Response Commission
AL Department of Environmental
Management
1751 Congressman W.L. Dickinson Drive
PO Box 301463
Montgomery, AL 36130-1463
Phone: (334) 260-2717
Section(s): 302, 304, 311, 312

Alaska
AK SERC Website:
http://www.ak-prepared.com
/serchome.htm/

Major General Jake Lestenkof, Co-Chair
Department of Military and Veteran Affairs
PO Box 5800
Fort Richardson, AK 99505-5800
Phone: (907) 428-6003
Section(s): None

Commissioner Michele Brown, Co-Chair
AK Department of Environmental
Conservation
410 Willoughby Avenue, Suite 105
Juneau, AK 99801-1795
Phone: (907) 465-5065 Section(s): None

Camille Stephens
AK Department of Environmental
Conservation
Spill Prevention and Response
410 Willoughby Avenue, Suite 105
Juneau, AK 99801-1795
Phone: (907) 465-5220
Section(s): 304, 311, 312
Website:
http://www.state.ak.us/local/akpages
/ENV.CONSERV/dspar/dec_dspr.htm#perp/

American Samoa
Faamausili Pola
Territorial Emergency Mgmt. Coordinating
Officer
American Samoan Government
Department of Public Safety
PO Box 1086
Pago Pago, AS 96799
Phone: (684) 633-1111
Section(s): 302, 304

Pati Faiai
American Samoa Environmental Protection
Agency
Office of the Governor
Pago Pago, AS 96799
Phone: (684) 633-2304
Section(s): 311, 312

Arizona

Michael P. Austin, Chairman
AZ Emergency Response Commission
5636 East McDowell Road
Phoenix, AZ 85008
Phone: (602) 231-6245
Fax: (602) 231-6313
Section(s): None

Daniel Roe, Executive Director
AZ Emergency Response Commission
5636 East McDowell Road
Phoenix, AZ 85008
Phone: (602) 231-6345
Fax: (602) 231-6313
E-mail: #AZSERC@DEM.STATE.AZ.US
Website:
http://www.state.az.us/es/azserc.htm/
Section(s): 302, 304, 311, 312

Arkansas

Joe Dillerd, Director
Office of Emergency Services
PO Box 758
Conway, AR 72033-0758
Section(s): None

Robert Johns
Office of Hazardous Materials Emergency
Management
PO Box 758
1835 South Donaughey
Conway, AR 72032
Phone: (501) 730-9789
Section(s): 302, 304, 311, 312

California

Dr. Richard Andrews, Director
Steve DeMello
Chemical Emergency Planning & Response
Commission
Office of Emergency Services
Hazardous Materials Division
2800 Meadowview Road
Sacramento, CA 95832
Phone: (916) 464-3281
Section(s): 302, 304, 311, 312

Colorado

Steve Gunderson
CO Emergency Planning Commission
CO Department of Health
Mail Code OE-EMU-B2
4300 Cherry Creek Drive South
Denver, CO 80222-1530
Phone: (303) 692-3022
Section(s): 302, 304, 311, 312

Connecticut

Gerard P. Goudreau, Chairman
State Emergency Response Commission
c/o Ulbrich Stainless Steels
1 Dudley Avenue
PO Box 610
Wallingford, CT 06492
Phone: (203) 269-2507, ext. 275
Section(s): None

Joseph B. Pulaski, SERC Administrator
State Emergency Response Commission
Department of Environmental Protection
79 Elm Street, 4th Floor
Hartford, CT 06106-5127
Phone: (860) 424-3373
Section(s): 302, 304, 311, 312

CT Office of Emergency Management
360 Broad Street
Hartford, CT 06105
Phone: (860) 566-3377
Fax: (860) 247-0664
Website:
http://www.state.ct.us/dps/ctoem.htm/

Delaware

Karen Johnson, Chair/Secretary
DE Department of Public Safety
PO Box 818
Dover, DE 19901
Phone: (302) 739-4321

Joe Wessels, SARA Title III Coordinator
DE Emergency Management Agency
PO Box 527
Delaware City, DE 19706
Phone: (302) 834-4531
General Info: (302) 326-6000
Toll Free (In-State Only): (800) 480-SERC
Section(s): 302
Website:
http://www.state.de.us/govern/agencies/
pubsafe/dema/indxdema.htm/

Andrea Maucher
Division of Air and Waste Management
Dept. of Natural Resources &
Environmental Control
89 Kings Highway
PO Box 1401
Dover, DE 19903
Phone: (302) 739-4791
Section(s): 304, 311, 312

District of Columbia
Samuel Jordan, Acting Director
Office of Emergency Preparedness
2000 14th Street, NW
Washington, DC 20009
Section(s): None

Michele Penick
Emergency Response Commission for Title
III
Office of Emergency Preparedness
2000 14th Street, NW
8th Floor
Washington, DC 20009
Phone: (202) 673-2101 Ext. 3159
Section(s): 302, 304, 311, 312

Florida
James F. Murley, Chair
FL Department of Community Affairs
Division of Emergency Management
State Emergency Response Commission
2555 Shumard Oak Boulevard
Tallahassee, FL 32399-2100
Phone: (904) 413-9970
Section(s): None

Florida (Continued)
Joseph Myers
Alternate SERC Chair
Division of Emergency Management
2555 Shumard Oak Boulevard
Tallahassee, FL 32399-2100
http://www.state.fl.us/comaff/DEM/index.ht
m/
Section(s): None

Eve Rainey, Compliance Planning
Florida Department of Community Affairs
Division of Emergency Management
State Emergency Response Commission
2555 Shumard Oak Boulevard
Tallahassee, FL 32399-2149
Phone: (904) 413-9970
Toll Free: (800) 635-7179
Section(s): 302, 304, 311, 312

Georgia
Joe Tanner, Chairman/Commissioner
GA Emergency Response Commission
205 Butler Street, S.E.
Suite 1252
Atlanta, GA 30334-4910
Phone: (404) 656-3500
Section(s): None

Burt Langley, Chief
GA Emergency Response Commission
205 Butler Street, S.E.
Floyd Tower East, Suite 1166
Atlanta, GA 30334-4910
Phone: (404) 656-6905
Section(s): 302, 304, 311, 312

GA Emergency Management Agency
PO Box 18055
Atlanta, GA 30316-0055
Phone: (404) 635-7000
Toll Free: (800) TRY-GEMA
Fax: (404) 635-7205
Audio News Service: (888) 216-0760
E-mail: webmaster@gema.state.ga.us
Website: http://www.State.Ga.US/GEMA/

Multimedia Environmental Compliance Guide for Food Processors

Guam
Jesus Salas, Administrator
Guam EPA
22439 GMF
Barrigada, GU 96921
Phone: (671) 475-1669
Section(s): 302, 304, 311, 312

Hawaii
Dr. Bruce S. Anderson
HI State Emergency Response
Commission
HI State Department of Health
PO Box 3378
Honolulu, HI 96801
Phone: (808) 586-4424
Section(s): None

Marsha Mealey, HEPCRA Coordinator
HI State Emergency Response
Commission
HI State Department of Health
PO Box 3378
919 Ala Moana Boulevard, Room 206
Honolulu, HI 96814
Phone: (808) 586-4694
E-mail: heer@eha.health.state.hi.us
Section(s): 302, 304, 311, 312

Idaho
Major General John Kane, Chairman
ID Bureau of Hazardous Materials
4040 Guard Street, Building 600
Gowen Field
PO Box 83720
Boise, ID 83720-3401
Phone: (208) 334-3263
Section(s): None
Website:
http://www.state.id.us/serc/index.html/

Bill Bishop, Director
ID Bureau of Hazardous Materials
4040 Guard Street, Building 600
Gowen Field
PO Box 83720
Boise, ID 83720-3401
Phone: (208) 334-3263
Section(s): 302, 304, 311, 312

Illinois
John Mitchell, Chairperson
State Emergency Response Commission
IL Emergency Management Agency
110 East Adams Street
Springfield, IL 62701-9963
Phone: (217) 782-2700
Section(s): None
Website: http://www.state.il.us/iema/

Oran Robinson, Supervisor
Hazardous Materials Compliance and
Enforcement
State Emergency Response Commission
c/o IL Emergency Management Agency
110 East Adams Street
Springfield, IL 62706-9963
Phone: (217) 782-4694
Section(s): 302, 304, 311, 312

Indiana
Patrick R. Ralston, Chairperson
State Emergency Response Commission
IN Government Center South
302 West Washington Street, Room E208
Indianapolis, IN 46204
Phone: (317) 232-3986
Section(s): None
Website: http://www.ai.org/ierc/

Cindi Wagner or Tom Madix
Indiana Department of Environmental
Management
2525 North Shadeland
PO Box 7024
Indianapolis, IN 46207
Phone: (317) 308-3039
Section(s): 302, 304, 311, 312

Iowa
William Zitterich, P.E., Director
IO Emergency Response Commission
IA DOT
Office of Maintenance Services
800 Lincoln Way
Ames, IA 50010
Phone: (515) 239-1396
Section(s): None

Walter Johnson, Vice-Chair
IO Emergency Response Commission
IO Division of Labor
1000 East Grand Avenue
Des Moines, IA 50319
Phone: (515) 281-8460
Section(s): None

Paul Sadler
Program Planner
Emergency Management Division
Hoover State Office Building, Level A
Des Moines, IA 50319
Phone: (515) 242-5171
E-mail: dsanders@max.state.ia.us
Section(s): 302

Pete Hamlin, Chief
Air Quality Bureau
IO Department of Natural Resources
7900 Hickman Road
Urbandale, IA 50322
Phone: (515) 281-8852
Section(s): 304

Anne Jackson
Sr. Industrial Hygenist
IO Division of Labor
1000 East Grand Avenue
Des Moines, IA 50319
Phone: (515) 281-8460
Section(s): 311, 312

Kansas

Bob Barid, SERC Chairman
General Motors
3201 Fairfax Traffic Way
Kansas City, KS 66115
Phone: (913) 573-7303
Section(s): None

Jon Flint, Section Chief
KS Emergency Response Commission
Right-to-Know Program
J Street and 2 North
Forbes Field Building 283
Topeka, KS 66620
Phone: (913) 296-1690
Section(s): 302, 311, 312

Frank Moussa
KS Division of Emergency Management
State Defense Building
2800 South West Topeka Boulevard
Topeka, KS 66611-1287
Phone: (913) 274-1409
E-mail: frankm@agtop.wpo.state.ks.us
Website: http://www.ink.org/public/kdem/
Section(s): 304

Kentucky

Ron Padgett, Executive Director
KY Emergency Response Commission
KY Disaster and Emergency Services
Boone National Guard Center
Frankfort, KY 40601-6168
Phone: (502) 564-8681
E-mail: rpadgett@kydes.dma.state.ky.us
Website: http://www.state.ky.us/agencies
/military/des.htm/
Section(s): None

Lucille Orlando, Branch Mgr. Technological
Hazards
KY Emergency Response Commission
KY Disaster and Emergency Services
Boone National Guard Center
Frankfort, KY 40601-6168
Phone: (502) 564-5223
Section(s): 302, 304, 311, 312

Louisiana

Capt. Mark S. Oxley, Chairman
LA Emergency Response Commission
Office of State Police, Right-to-Know Unit
7901 Independence Boulevard
Building A
Baton Rouge, LA 70806
Phone: (504) 925-6113
Section(s): None

Robert Hayes, Management Analyst
LA Emergency Response Commission
Office of State Police, Right-to-Know Unit
7901 Independence Boulevard
Building A
Baton Rouge, LA 70806
Phone: (504) 925-6113
Section(s): 302, 304, 311, 312

Linda Brown, TRI Coordinator
Department of Environmental Quality
Office of the Secretary
PO Box 82263
Baton Rouge,LA 70884-2263
Phone: (504) 765-0737
Section(s): 313

LA Office of Emergency Preparedness
Website: http://199.188.3.91/chemihaz.htm/

Maine
John W. Libby, Chairman
State Emergency Response Commission
ME Emergency Management Agency
72 State House Station
Augusta, ME 04333
Phone: (207) 287-4080
Toll Free: (800) 452-8735
Fax: (207) 287-4079
E-mail: john.w.libby@state.me.us
Website: http://www.state.me.us/mema/
Section(s): None

Rayna Leibowitz
State Emergency Response Commission
72 State House Station
Augusta, ME 04333
Phone: (207) 287-4080
Toll Free: (800) 452-8735
Fax: (207) 287-4079
E-mail: rayna.b.leibowitz@state.me.us
Section: 302, 304, 311, 312

Maryland
David McMillian, Chairman
Governor's Emergency Management
Advisory Council
c/o MD Emergency Management Agency
2 Sudbrook Lane East
Pikesville, MD 21208
Phone: (410) 486-4422
Section(s): None
Website: http://www.mema.state.md.us/

Patricia Williams, Environmental Specialist
MD Department of the Environment
Technical and Regulatory Services
Administration
Community Right-to-Know Section
2500 Broening Highway
Baltimore, MD 21224
Phone: (410) 631-3800
Section(s): 302, 304, 311, 312

Massachusetts
Kathleen O'Toole, Chair
Secretary of Public Safety
MA Emergency Response Commission
Executive Office of Public Safety
One Asburton Place, Room 2133
Boston, MA 02108
Phone: (617) 727-7725
Section(s): None

John Tommaney
MA Emergency Management Agency
PO Box 1496
400 Worcester Road
Framingham, MA 01701-0317
Phone: (508) 820-2000
Website:
http://www.magnet.state.ma.us/mema
/homepage.htm/
Section(s): 302, 304, 311, 312

Michigan

Captain Robert Tarrant, Chairperson
State Emergency Response Commission
EMD/MI Department of State Police
4000 Collins Road
PO Box 30636
Lansing, MI 48909-8136
Phone: (517) 333-5041
Website:
http://www.msp.state.mi.us/division
/emd/emdweb1.htm/
Section(s): None

Robert Jackson, Chief, Grants & Information
State Emergency Response Commission
MI Department of Environmental Quality
Environmental Assistance Division, Title III
Office
300 South Washington
PO Box 30457
Lansing, MI 48909
Phone: (517) 373-2731
Section(s): 302, 304, 311, 312

Minnesota

Dennis Shershen, Chairperson
State Emergency Response Commission
Truth Hardware Corporation
700 West Bridge Street
Owatonna, MN 55060
Phone: (507) 444-4481
Section(s): None

Paul Aasen
MN Emergency Response Commission
State Capitol Building, Room B-5
75 Constitution Avenue
St. Paul, MN 55155
Phone: (612) 282-5391
Website:
http://www.dps.state.mn.us/emermgt/
Section(s): 302, 304, 311, 312

Mississippi

J.E. Maher, Chairman
MS Emergency Response Commission
MS Emergency Management Agency
PO Box 4501
Jackson, MS 39296-4501
Phone: (601) 960-9000
Section(s): None

John David Burns
MS Emergency Response Commission
MS Emergency Management Agency
PO Box 4501
Jackson, MS 39296-4501
Phone: (601) 960-9000
Section(s): 302, 304, 311, 312

Chuck Carter
Plan Operations Manager
MS Emergency Response Commission
MS Emergency Management Agency
PO Box 4501
Jackson, MS 39296-4501
Phone: (601) 960-9000
Section(s): 302, 304, 311, 312

Missouri

Bob Dopp, Executive Chief
MS Emergency Response Commission
PO Box 3133
Jefferson City, MO 65102
Phone: (573) 526-9237
E-mail: bdoppol@mail.state.mo.us
Website: http://www.sema.state.mo.us
/mercc.htm/
Section(s): 302, 304, 311, 312

MS Emergency Preparedness Association
Website: http://www.iland.net/mepa/

Montana

Tom Ellerhoff, Chairman
MT Emergency Response Commission
ESD/DHES, Metcalf Building
1520 E. 6th Avenue
PO Box 200901
Helena, MT 59620-0901
Phone: (406) 444-5263
Section(s): 311, 312

Jim Greene
Disaster Emergency Services
1100 North Main
PO Box 4789
Helena, MT 59604-4789
Phone: (406) 444-6911
Section(s): 302, 304

Nebraska
Major General Stanley M. Heng, Chairman
NE Emergency Response Commission
1300 Military Road
Lincoln, NE 68508-1090
Phone: (402) 471-7100
Section(s): None

Dale Bush, Interim Coordinator
NE Department of Environmental Quality
PO Box 98922
State House Station
Lincoln, NE 68509-8922
Phone: (402) 471-4237
Section(s): 304, 311, 312

Dave Kudmore
NE Emergency Management Agency
1300 Military Road
Lincoln, NE 68508-1090
Phone: (402) 471-7420
Website:
http://www.sarpy.com/ema/index.htm/
Section(s): 302

Nevada
Marvin Carr
State Emergency Response Commission
555 Wright Way
Carson City, NV 89711-0925
Phone: (702) 687-6973
E-mail: sercinfo@govmail.state.nv.us
Section(s): 302

Division of Emergency Management
2525 South Carson Street
Carson City, NV 89710
Phone: (702) 687-4240
Section(s): 304

Larry W. Bennett
SERC Co-Chair
Southern Pacific Transportation Co.
50 Washington Street
Suite 101
Reno, NV 89530
Phone: (702) 323-3688
Section(s): 302, 304, 311, 312

Karen Larson
Co-chair of SERC
Clark County Manager's Office
500 South Grand Central Parkway
Las Vegas, NV 89155-1111
Phone: (702) 455-6186
Section(s): 302, 304, 311, 312

New Hampshire
George L. Iverson, Director
NH Office of Emergency Management
Title III Program
State Office Park South
107 Pleasant Street
Concord, NH 03301
Phone: (603) 271-2231
Website: http://www.nhoem.state.nh.us
/nhoem.ssi/
Section(s): None

Leland Kimball
NH State Emergency Management Agency
Title III Program
State Office Park South
107 Pleasant Street
Concord, NH 03301-3809
Phone: (603) 271-2231
Section(s): 302, 304, 311, 312

New Jersey
Shirlee Schiffman, Chief
NJ Dept. of Environmental Protection
Bureau of Chemical Release Information
and Prevention
Health and Analytical Programs
401 East State Street, CN 405
Trenton, NJ 08625
Phone: (609) 984-3219
Section(s): 302

Stan Delikat
Director of Responsible Party Site
Remediation
Bureau of Emergency Response
401 East State Street CN-028
Trenton, NJ 08625
Phone: (609) 633-2168
Section(s): 304

Alan Bookman
NJ Dept. of Environmental Protection and
Energy
Bureau of Chemical Release Information
and Prevention
Health and Analytical Programs
401 East State Street CN-405
Trenton, NJ 08625
Phone: (609) 984-5338
Section(s): 311, 312

NJ Hazardous Materials Advisory Council
Website: http://www.hmac-inc.org/

New Mexico
Ray Denison, Chairman
NM Emergency Response Commission
NM Department of Public Safety
PO Box 1628
Santa Fe, NM 87504-1628
Phone: (505) 827-3376
Section(s): None

Max Johnson, Coordinator
NM Emergency Response Commission
Technological Hazards Bureau
Emergency Management
PO Box 1628
Santa Fe, NM 87504-1628
Phone: (505) 476-9620
Section(s): 302, 304, 311, 312

New York
Major General John Fenimore, Vice
Chairman
NY Emergency Response Commission
State Emergency Management Office
Building 22, Suite 101
1220 Washington Avenue
Albany, NY 12226-2251
Phone: (518) 457-2222
Website:
http://www.nysemo.state.ny.us/SERC
/serc.html
Section(s): None

NY Emergency Response Commission
NY State Dept. of Environmental
Conservation
Bureau of Spill Prevention and Response
50 Wolf Road
Room 340
Albany, NY 12233-3510
Phone: (518) 457-4107
Section(s): 302, 304, 311, 312

North Carolina
Billy Camaron, Chairman
NC Emergency Response Commission
NC Division of Emergency Management
116 West Jones Street
Raleigh, NC 27603-1335
Phone: (919) 733-3825
Website: http://www.dem.dcc.state.nc.us/
Section(s): None

Esther Castaldo
NC Emergency Response Commission
NC Division of Emergency Management
116 West Jones Street
Raleigh, NC 27603-1335
Phone: (919) 733-3899
Section(s): 302, 304, 311, 312

NC Emergency Management Association
Website:
http://www.mindspring.com/~dobesar
/ncema.htm/

North Dakota
Lyle Gallagher, Director of Communications
State Radio Communications Department
PO Box 5511
Bismarck, ND 58506
Phone: (701) 328-2121
E-mail: lgallagh@pioneer.state.nd.us
Section(s): 304

Douglas C. Friez, Chairman
ND State Division of Emergency
Management
PO Box 5511
Bismarck, ND 58506-5511
Phone: (701) 328-3300
E-mail: msmail.doug@ranch.state.nd.us
Website: http://pioneer.state.nd.us/dem/
Section(s): 302, 311, 312

Northern Mariana Islands
Felix Sasamoto, Civil Defense Coordinator
Office of the Governor
Capitol Hill
Commonwealth of Northern Mariana Islands
Saipan, MP 96950
Phone: (011) (670) 322-9529
Section(s): 302, 304, 311, 312

Ohio
Dale Shipley, Chairman
State Emergency Response Commission
OH Emergency Management Agency
2855 W. Dublin-Granville Road
Columbus, OH 43235-2206
Phone: (614) 889-7150
Website:
http://www.ohio.gov/ODPS/division
/ema/index.htm/
Section(s): None

Ken Schultz, Co-Chairperson
State Emergency Response Commission
OH Environmental Protection Agency
PO Box 163669
Columbus, OH 43216-3669
Phone: (614) 644-2081
Section(s): None

Jeff Beattie
OH Emergency Response Commission
OH Environmental Protection Agency
Office of Emergency Response
1800 Watermark Drive
Columbus, OH 43215-1099
Phone: (614) 644-2081
Section(s): 302, 304, 311, 312

OH Chemical Emergency and
Preparedness Section
Website: http://www.epa.ohio.gov/derr
/derrmain.html/

Oklahoma
Monty Elder
Department of Environmental Quality
Customer Services
1000 North East Tenth Street
Oklahoma City, OK 73117-1212
Phone: (405) 271-1400 Ext. 192
E-mail: monty.elder@OKLAOSF.state.ok.us
Section(s): 302, 311, 312

Lynne Moss
Department of Environmental Quality
1000 N.E. 10th Street
Oklahoma City, OK 73117-1212
Phone: (405) 271-7363
E-mail: lynne.moss@OKLAOSF.state.ok.us
Section(s): 304

OK Department of Civil Emergency
Management
Website: http://www.onenet.net/~odcem
/index.html/

Oregon
James Mazza, Hazardous Materials
Planning Coordinator
c/o Department of Oregon State Police
Oregon Emergency Management
595 Cottage Street, NE
Salem, OR 97310
Phone: (503) 378-2911
Section(s): None

Bob Albers
OR Emergency Response Commission
c/o State Fire Marshall
4760 Portland Road, NE
Salem, OR 97305-1760
Phone: (503) 378-3473 Ext. 262
E-mail: bob.albers@state.or.us
Section(s): 302, 304, 311, 312

Pennsylvania
Lt. Governor Mark S. Schweiker, Chairman
PA Emergency Management Council
PO Box 3321
Harrisburg, PA 17105-3321
Phone: (717) 651-2001
Section(s): None

Robert F. Broyles, Chief
Chemical Emergency Preparedness
Division
PA Emergency Management Agency
PO Box 3321
Harrisburg, PA 17105-3321
Phone: (717) 651-2199
Section(s): 302, 304

Thomas Ward or Lynn Snead
PA Emergency Management Council
Bureau of Worker and Community
Right-to-Know
Room 1503
Labor and Industry Building
7th and Forster Streets
Harrisburg, PA 17120
Phone: (717) 783-2071
Section(s): 311, 312

PA Emergency Management Agency
Website: http://www.pema.state.pa.us/

Puerto Rico
Hector Russe Martinez, Chairman
PR Emergency Response Commission
Environmental Quality Board
PO Box 11488
Santurce, PR 00910
Phone: (787) 767-8056
Section(s): None

Genaro Torres
Director of Superfund and Emergency
Division
Title III - SARA Section 313
PR Environmental Quality Board
Sernades Junco Station
PO Box 11488
Santurce, PR 00910
Phone: (787) 767-8056
Section(s): 302, 304, 311, 312

Rhode Island
Raymond LaBelle, Executive Director
RI Emergency Response Commission
645 New London Avenue
Cranston, RI 02920-3003
Phone: (401) 222-3039
Section(s): None

John Aucott
RI Emergency Response Commission
645 New London Avenue
Cranston, RI 02920-3003
Phone: (401) 946-9996
E-mail: aucott@mindspring.com
Section(s): 302, 304

Patrice Carvaretta
RI Department of Labor
Division of Occupational Safety
610 Manton Avenue
Providence, RI 02909
Phone: (401) 457-1829
Section(s): 311, 312

South Carolina
Stan M. McKinney, Chairman
SC Emergency Response Commission
c/o Emergency Preparedness Division
1429 Senate Street
Columbia, SC 29201
Phone: (803) 734-8020
Section(s): None

Michael Juras
SC State Emergency Response Comm.
EPCRA Reporting Point
SC Dept. of Health & Environmental Control
2600 Bull Street
Columbia, SC 29201
Phone: (803) 896-4117
Section(s): 304

Peter Saussy
SC State Emergency Response
Commission
SC Dept. of Health and Environmental
Control
2600 Bull Street
Columbia, SC 29201
Phone: (803) 896-4116
Section(s): 311, 312

John Berzins
SC Emergency Response Commission
c/o Emergency Preparedness Division
1429 Senate Street
Columbia, SC 29201
Phone: (803) 734-8020
Section(s): 302

SC Emergency Management Agency
Website: http://www.state.sc.us/epd/

South Dakota
Bob McGrath
SD Emergency Response Commission
Department of Environment & Natural
Resources
Joe Foss Building
523 East Capitol
Pierre, SD 57501-3181
Phone: (605) 773-3296
Section(s): None

Lee Ann Smith, Title III Coordinator
SD Emergency Response Commission
Department of Environment & Natural
Resources
Joe Foss Building
523 East Capitol
Pierre, SD 57501-3181
Phone: (605) 773-3296
Section(s): 302, 304, 311, 312

SD Division of Emergency Management
Website:
http://www.state.sd.us/state/executive
/military/sddem.htm/

Tennessee
John White, Chairman
TN Emergency Response Commission
TN Emergency Management Agency
3041 Sidco Drive
Nashville, TN 37204
Phone: (615) 741-0001
Section(s): None

Betty Eaves, Director
TN Emergency Management Agency
3041 Sidco Drive
Nashville, TN 37204
Phone: (615) 741-2986
Section(s): 302, 304, 311, 312
Website: http://www.state.tn.us/military
/tema.html/

Texas
Dudley Thomas, Director
TX Emergency Response Commission
PO Box 4087
Austin, TX 78773-0001
Phone: (512) 424-2429
Section(s): None

Tom Millwee, Co-Chair
TX Emergency Response Commission
Division of Emergency Management
PO Box 4087
Austin, TX 78773-0001
Phone: (512) 424-2429
Section(s): None

Annabelle Dillard
TX Department of Health
Hazard Communication Branch
1100 West 49th Street
Austin, TX 78756
Phone: (512) 834-6603
Toll Free: (800) 452-2791
E-mail: adillard@beh.tdh.state.tx.us
Section(s): 302, 311, 312

David Barker
Emergency Response Team (MC 142)
TX Natural Resource Conservation
Commission
Room 241
PO Box 13087
Austin, TX 78711
Phone: (512) 463-7727
Section(s): 304

TX Division of Emergency Management
Website:
http://isadore.tsl.state.tx.us/tx/DEM/

Utah
Lorayne Frank, Director, Co-Chairperson
Department of Public Safety
Division of Comprehensive Emergency
Management
1110 State Office Building
Salt Lake City, UT 84114
E-mail: pscm.loraynefrank@state.ut.us
Phone: (801) 538-3400
Section(s): None

Jerry Nortin
Division of Comprehensive Emergency
Management
1110 State Office Building
Salt Lake City, UT 84114
Phone: (801) 538-3774
E-mail: pscm.jerrynortin@state.ut.us
Section(s): 302, 304

Neil Taylor
State Emergency Response Commission
UT Department of Environmental Quality
Div. of Environmental Response &
Remediation
PO Box 144840
168 North 1950 West, 1st Floor
Salt Lake City, UT 84116
Phone: (801) 536-4102
E-mail: ntaylor@deq.state.ut.us
Website:
http://www.eq.state.ut.us/eqerr/serc
/SERCHOME.HTM
Section(s): 311, 312

Vermont
George Lowe, Interim Chair
Department of Public Safety
103 South Main Street
Waterbury, VT 05676
Phone: (802) 244-8721
Section(s): 304

Randy Bronson
Department of Public Safety
103 S. Main Street
Waterbury, VT 05676
Phone: (802) 244-8721
Section(s): 302, 311, 312

Virgin Islands
Austin Moorehead
Dept. of Planning & Natural Resources
Division of Environmental Protection
Watergut Homes
1118 Christiansteb
St. Croix, VI 00820-5065
Phone: (809) 773-0565
Section(s): 302, 304, 311, 312

Virginia
Chairman
VA Emergency Response Commission
VA Department of Emergency Services
310 Turner Road
Richmond, VA 23225
Section(s): None

Paul Spaulding
VA Emergency Response Council
c/o VA Department of Environmental Quality
PO Box 10009
Richmond, VA 23240-0009
Phone: (804) 698-4480 or 4489
E-mail: plspaulding@deq.state.va.us
Section(s): 302, 304, 311, 312

VA Department of Emergency Services
Website: http://www.vdes.state.va.us/

Washington

Chief Robert Johnson
Auburn Fire Department
1101 D Street NE
Auburn, WA 98002
Phone: (206) 931-3060
Section(s): None

John Ridgway
CRTK Unit
Dept. of Ecology
PO Box 47659
Olympia, WA 98504-7659
Phone: (360) 407-6713 or (800) 633-7585
E-mail: jrid461@ecy.wa.gov
Section(s): 302, 304, 311, 312

Sadie Whitener
CRTK Unit
Dept. of Ecology
PO Box 47659
Olympia, WA 98504-7659
Phone: (360) 407-6729, or (800) 633-7585
E-mail: swhit461@ecy.wa.gov
Section(s): 302, 304, 311, 312

WA State Emergency Management Division
Website: http://www.wa.gov/mil/wsem/

West Virginia

Carl L. Bradford, Director
WV Emergency Response Commission
WV Office of Emergency Services
Main Capitol Building 1
Room EB-80
Charleston, WV 25305-0360
Phone: (304) 558-5380
Section(s): 302, 304, 311, 312

Wisconsin

Steven D. Sell, Chairperson
State Emergency Response Board
Division of Emergency Management
PO Box 7865
2400 Wright Street
Madison, WI 53707-7865
Phone: (608) 242-3232
Section(s): None

William Clare
State Emergency Response Board
2400 Wright Street
Madison, WI 53707-7865
Phone: (608) 242-3232
Website:
http://badger.state.wi.us/agencies/dma
/wem/serb.htm/
Section(s): 302, 304, 311, 312

Wyoming

Pete Illoway, Chair
State Emergency Response Commission
Wyoming Emergency Management Agency
5500 Bishop Boulevard
Cheyenne, WY 82009-3320
Phone: (307) 777-4900
Section(s): None

Bob Bezek
Wyoming Emergency Response
Commission
Wyoming Emergency Management Agency
Department of Environmental Quality
PO Box 1709
5500 Bishop Boulevard
Cheyenne, WY 82009-3320
Phone: (307) 777-4900
Section(s): 302, 304, 311, 312

Local Emergency Planning Committee (LEPC) Websites

To obtain information on your LEPC, contact your State Emergency Response Commission (SERC) or state environmental agency, or visit the following websites:

Alabama
Madison County EPC (MCEPC)
http://www.ci.huntsville.al.us/ema/

Coffee County LEPC
http://www.snowhill.com/~twilkins/cclepc.html/

Alaska
Alaska LEPC Chairs
http://www.ak-prepared.com/lepclist.htm/

Colorado
Araphoe County EPC
http://www.sni.net/arapc/

Jefferson County EPC
http://www.gablehouse-epel.com/Jeffco1.htm/

Larimer County LEPC
http://www.co.larimer.co.us/depts/sherif/emerg6.htm/

Connecticut
Connecticut LEPCs by Map
http://www.lepc-len.org/

Florida
Apalachee Regional Planning Council
http://www.thearpc.org/emermgt/index.html/

Central Florida Regional Planning Council
http://lakeland.tsolv.com/~cfrpc/

District I LEPC
http://www.wfrpc.dst.fl.us/wfrpc/lepc.htm/

East Central Florida Regional Planning Council
http://www.orlinter.com/ecfrpc/const.htm/

Florida District X LEPC
http://www.tcrpc.org/lepcndx.htm/

North Central Florida LEPC
http://www.afn.org/~lepc/

South Florida LEPC
http://www.sfrpc.com/current/menu522.htm/

Southwest Florida Regional Planning Council
http://www.swfrpc.org/haz.htm/

Tampa Bay Regional Planning Council
http://access.tampabayrpc.org/lepc/lepc.htm/

District 5 LEPC (Withlacoochee Planning Council)
http://www.atlantic.net/~wrpc/lepc.htm/

Illinois
Kane County EPC
http://www.ameritech.net/users/kcema/kcema1.htm/

Indiana
Hamilton County LEPC
http://www.members.iquest.net/~lepc/index.htm/

Tippecanoe County LEPC
http://www.nfe.com/tema/lepc/lepc.htm/

Evansville LEPC
http://www.evansville.net/eville/services/lepc.html/

Indiana LEPCs by Map
http://www.ai.org/ierc/html/inmap.html/

Kansas
Labette County LEPC
http://www.geocities.com/Heartland/Plains/3
524/lepc.html/

Henderson County LEPC
http://www.dynasty.net/users/
hendersonlepc/

Nelson County Disaster and Emergency
Services
http://www.bardstown.com/~des760/

Louisiana
List of LA LEPCs & Chairpersons
http://www.dps.state.la.us/lsp/tess/rtk
/98RENEWL.html/

Beauregard Parish Office of Emergency
Preparedness
http://www.beau.lib.la.us/~oem/

Ouachita Parish LEPC
http://www.bayou.com:80/~civil/

Iberville Parish LEPC
http://www.parish.iberville.la.us/911/icaer.ht
m/

Maine
Oxford County LEPC
http://www.megalink.net/~oxctyema/lepc.
htm/

Piscataquis County LEPC
http://www.kynd.com/~pcema/lepc.htm/

Maryland
Harford County LEPC
http://www.co.ha.md.us/lepc.html/

Massachusetts
MA List of LEPCs
http://www.magnet.state.ma.us/dep/bwp/dh
m/tura/turapubs.htm/

Northern Middlesex Area EPC
http://www.tiac.net/users/mclatchy/nmaepc
/nmaepc.htm/

Merrimack Valley Regional LEPC
http://www.shore.net/~mvrlepc/

City of Peabody LEPC
http://www.shore.net/~chief/lepc1.htm/

Michigan
List of MI LEPCs
http://www.deq.state.mi.us/ead/sara/

Ingham County LEPC
http://www.aware.msu.edu/

Jackson County LEPC
http://www.co.jackson.mi.us/ems/lepc.htm/

Monroe County LEPC
http://www.monroe.lib.mi.us/mcemd/lepc.
htm/

Washtenaw County LEPC
http://www.co.washtenaw.mi.us/DEPTS/eis
/eisresp.htm/

Missouri
Mid-America LEPC
http://www.marc.org/lepc.htm/

St. Charles County LEPC
http://www.win.org/county/depts/emergenc
/sccg275h.htm/

Missouri LEPCs by Map
http://www.iland.net/mepa/lepc.html/

Cooper County LEPC
http://mo-river.net/community/ema/cooper/
lepc/lepc.htm/

St. Louis County LEPC
http://www.co.st-louis.mo.us/oem/lepc.HTM/

Nebraska
Sarpy County LEPC
http://www.sarpy.com/ema/lepc.htm/

New Jersey
Fair Lawn LEPC

http://hannover.park.org/Guests/Fairlawn/
em.html/

New Mexico
Eddy County LEPC
http://www.carlsbadnm.com/emerprep/lepc.
htm/

New York
Schenectady County, NY LEPC
http://ns1.crisny.org/government/capreg/
sclepc/index.html/

Oswego LEPC
http://www.co.oswego.ny.us/info/newslet/lep
c.html/

Erie County LEPC
http://www.buffalo.edu/esi/eclepc.htm/

North Carolina
Rowan County LEPC
http://www.co.rowan.nc.us/es/lepc/
lepc.htm/

Wake County LEPC
http://www.state.nc.us/Wake/depts/PubSafe
/major/lepc/lepc_hp.htm/

Davidson County LEPC
http://www.co.davidson.nc.us/emfire/lepc.
htm/

Ohio
Ohio LEPC
http://www.state.oh.us/odps/division/ema/
LEPC.htm/

Athen's County LEPC
http://www.seorf.ohiou.edu/~xx017/hide/lepc
/lepc/

Franklin County & Columbus LEPC
http://pie.mhsc.org/FCPHDB/envhlth/lepc00
0.htm/

Portage County LEPC
http://www.pcema.com/pclepc.html

Ohio State Emergency Coordinators
http://www.epa.ohio.gov/derr/cepps/cepd/le
pc.html/

Oregon
Oregon LEPC
http://members.aol.com/oregonlepc/lepc.
html/

Marion County Emergency Management
http://www.open.org/memanage/

Pennsylvania
Bucks County LEPC
http://www.buckslepc.com/

Berks County LEPC
http://www.snowhill.com/~twilkins/cclepc.
html/

Mifflin County LEPC
http://lcworkshop.com/mifflinco/lepc.htm/

Texas
Grayson County LEPC
http://members.tripod.com/~Burlena/index-
2. html/

Taylor County LEPC
http://www.fireweb.com/TaylorLEPC/

Houston LEPC
http://www.ci.houston.tx.us/lepc/index.html/

Fort Bend County LEPC
http://www.fortbend.lib.tx.us/emerman/lepc
/index.htm/

Cameron County LEPC/Comite Local de
Ayuda Mutua (CLAM)
http://www.triplesoft.com/bisn/trip/jlpre
s.html/

Matagorda County LEPC
http://www.man-net.org/emergency/lepc/lep
c.html/

Harris County: City of Baytown LEPC
http://www.ci.baytown.tx.us/dept/emer/lepc.
txt/

Harris County: City of Pasadena LEPC
http://www.flash.net/~pasalepc/

Virginia
Fairfax Joint LEPC
http://members.aol.com/Hazmt96/hmerp.htm/

Washington
Kings County LEPC
http://www.metrokc.gov/prepare/hazardou.htm/

Spokane City/County LEPC
ttp://www.spokanecounty.org/emergencymgmt/lepc.htm/

Wisconsin
Ozaukee County Department of Emergency Management
http://www.execpc.com/~n9unr/ozares/

Milwaukee County LEPC
http://www.mkesheriff.org/html/emerg/lepc.htm/

Wyoming
Wyoming Emergency Management Agency
http://132.133.10.9/

Appendix C. References

1. The Guide: What it Is; What it Does

Barlett, Kerry L., Lester, Richard R., and Pojasek, Robert B. *Prioritizing Pollution Prevention Opportunities With Activity-Based Costing.* Pollution Prevention Review. Cambridge: Cambridge Environmental Inc. October 1995.

U.S. Department of Commerce, Bureau of the Census. *1994 Census of Manufacturers: Subject Series.* Accessed via the Internet at http://www.census.gov/prod/www/titles.html#mm.

Office of Management and Budget, Executive Office of the President. *Standard Industrial Classification Manual,* 1987.

U.S. Environmental Protection Agency.

An Introduction to Environmental Accounting As A Business Management Tool: Key Concepts and Terms. EPA 742-R-95-001. June 1995.

Incentives for Self-Policing: Discovery, Disclosure, Correction and Prevention of Violations. Federal Register: December 22, 1995 (60 FR 246).

Policy on Compliance Incentives for Small Businesses. Federal Register: June 3, 1996 (61 FR 27984).

EPA Position Statement on Environmental Management Systems and ISO 14001 and a Request for Comments on the Nature of the Data to be Collected from Environmental Management System/ISO 14001 Pilots. 63 FR 12094-97. March 12, 1998

Statement on State Audit Privilege and Immunity Laws. Steven A. Herman. Assistant Administrator. Office of Enforcement And Compliance Assurance. Before the Committee on Commerce Oversight And Investigations Subcommittee. United States House of Representatives. Presented on March 17, 1998.

2. Guide to EPA's Major Federal Environmental Statutes

National Food Processors Association.

1993 Report of Scientific and Technical Activities. Washington, D.C. 1994

Source Book of State Laws & Regulations for Food Processors. June 1996.

U.S. Environmental Protection Agency.

FY 1991 Enforcement Accomplishments Report. Office of Enforcement. Office of Compliance Analysis and Program Operations. April 1992. EPA 300-R92-008.

FY 1992 Enforcement Accomplishments Report. Office of Enforcement. Office of Compliance Analysis and Program Operations. April 1993. EPA 300-R93-001.

FY 1993 Enforcement Accomplishments Report. USEPA. Office of Enforcement. Office of Compliance Analysis and Program Operations. April 1994. EPA 300-R-94-003.

Industry Profiles: Food and Kindred Products and Stone, Clay, Glass, and Concrete. Office of Solid Waste. July 1994.

FY 1994 Enforcement and Compliance Assurance Accomplishments Report. Office of Enforcement and Compliance Assurance. Office of Compliance. May 1995. EPA 300-R-95-004.

Preliminary Summary of Food Sector Cases From Enforcement and Compliance Assurance Accomplishments Reports Fiscal Years 1991-1994. Office of Compliance. 1995.

FY 1995 Enforcement and Compliance Assurance Accomplishments Report. Office of Enforcement and Compliance Assurance. Office of Compliance. July 1996. EPA 300-R-96-006.

FY 1996 Enforcement and Compliance Assurance Accomplishments Report. Office of Enforcement and Compliance Assurance. Office of Compliance. 1997.

Preliminary Summary of Food Sector Cases From Enforcement and Compliance Assurance Accomplishments Reports Fiscal Years 1995-1996. Office of Compliance. 1997.

Compliance and Enforcement History of the Food Processing Industry: December 1991 - December 1996 (Draft). Office of Compliance. April 1997.

FY 1997 Enforcement and Compliance Assurance Accomplishments Report. (Draft) Office of Enforcement and Compliance Assurance. Office of Compliance. April 1998.

Walsh, James L. and Ray, B. Russell. *The Impact of Environmental Regulations on the Food Processing Industry.* Accessed via the Internet at http://w\eoeml-www.gtri.gatech.edu/lab/eeb_foodpac-regs.html.

3. Understanding the Process: Inputs, Outputs, and Applicable Environmental Regulations

U.S. Environmental Protection Agency. *Industry Profiles: Food and Kindred Products and Stone, Clay, Glass, and Concrete.* Office of Solid Waste. Prepared by ICF Inc. July 1994.

Walsh, James L., Ross, Charles, C., and Valentine, G. Edward.

"Food Processing Waste." *Water Environment Research.* Vol. 65, No. 4. June 1993.

"Food Processing Waste." *Water Environment Research.* Vol. 66, No. 4. June 1994.

4. How Do I Comply With Wastewater Discharge and Related Regulations?

American Bakers Association. "Model Storm Water Permit." February 19, 1992.

Carawan, Roy et.al. "Spinoff On Fruit and Vegetable Water and Wastewater Management." *Industrial Water Conservation References of Food Processing.* California Department of Water Resources. 1989.

Law/Crandall, Inc. *Storm Water Pollution Prevention Plan (Draft).* Prepared for Martino's Bakery, Burbank, California. Law Project No. 58-162001.01. September 21, 1992.

National Canners Association. *Liquid Wastes From Processing Fruits, Vegetables, and Specialities.* Western Research Laboratory. Berkeley, California. # 1671, D-2765.

Robillard, Paul D. And Elliot, H.A. *Waste Management and Water Conservation in the Food Processing Industry. Proceedings of the Food Processing Waste Management and Water Conservation Conference.* November 1989.

U.S. Environmental Protection Agency.

Red Meat Processing: Segment of the Red Meat Product and Rendering Processing Point Source Category. Office of Air and Water Programs. EPA-440/1-74-012-a. February 1974.

Dairy Product Processing: Point Source Category. EPA-440/1-74-021-a. Office of Air and Water Programs. May 1974.

A Plain English Guide to the EPA Part 503 Biosolids Rule. Office of Wastewater Management. EPA/833/R-93/003. September 1994.

"Storm Water Discharges Associated With Industrial Activity from Food and Kindred Products Facilities." 60 FR 189. September 29, 1995.

Overview of the Storm Water Program. Office of Water. EPA 833-R-96-008. June 1996.

Water Pollution Control Federation. *Municipal Strategies for the Regulation of Sewer Use: Manual of Practice No. SM-7.* 1988.

5. How Do I Comply With Safe Drinking Water Regulations?

U.S. Environmental Protection Agency.

Environmental Pollution Control Alternatives: Drinking Water Treatment for Small Communities. April 1990.

Class V Injection Wells Regulatory Amendments. Office of Water. August 1995. Accessed via the Internet on May 5, 1997. http://www.site.net/gwpc/uicvepa.htm

6. How Do I Comply With Air Regulations?

Faulkner, Kelly F. "CAA's Operating Permit Program: Who's On First." *Environmental Solutions.* Vol. 7, No. 6. June 1994.

Food Manufacturing Coalition. *Odor Control of Food Processing Operations By Applications of Air Cleaning Technologies.* Project Code: ET-5-A (11). University of Nebraska Food Processing Center. Dr. Steve Taylor, Director. June 1, 1996. Accessed via the Internet at http://ceres.esusda.gov/fimc/need-08.htm

Government Institutes, Inc., *Clean Air Handbook 3rd Edition* 1998.

Parnell, C.B. and Wakelyn, P.J. *Regulation of Agricultural Operations Using Emission Factors and Process Weight Tables.*

Rocco, Vincent A. "Air Pollution Control." *Environmental Solutions.* Vol. 8, No. 6. June 1995.

U.S. Environmental Protection Agency.

Odor and Corrosion Control in Sanitary Sewerage Systems and Treatment Plants. Office of Research and Development. EPA/625/1-85-018. October 1985.

Asbestos/NESHAP Adequately Wet Guidance. Office of Air and Radiation. EPA 340/1-90-019. December 1990.

Asbestos NESHAP Milling, Manufacturing, and Fabricating Operations: Field Inspection Checklist. Office of Enforcement. March 1992.

Asbestos/NESHAP Regulated Asbestos Containing Materials Guidance. Office of Air and Radiation. EPA 340/1-90-018. December 1990.

Applicability of the Asbestos NESHAP to Asbestos Roofing Removal Operations: Guidance Manual. Office of Air and Radiation. EPA 340-B-94-001. August 1994.

Common Questions on the Asbestos NESHAP. Office of Air and Radiation. EPA 340/1-90-021. December 1990.

A Guide to the Asbestos NESHAP: As Revised November 1990. Office of Air and Radiation. EPA 340/1-90-015. November 1990.

Guidelines for Asbestos NESHAP Demolition and Removal Inspection Procedures. Office of Air and Radiation. EPA 340/1-90-007. November 1990.

Reporting and Recordkeeping Requirements for Waste Disposal: A Field Guide. Office of Air and Radiation. EPA 340/1-90-016. November 1990.

Risk Management Planning: Accidental Release Prevention. Final Rule: Clean Air Act Section 112(r) Fact Sheet. EPA 550-F-96-002. May 1996.

General Guidance for Risk Management Programs (40 CFR 68). EPA 550-B-98-003. July 1998.

RMP Series. EPA's Risk Management Program: How Does It Affect Operators of Ammonia Refrigeration Systems? EPA 550-F98-006. 1998

RMP Series. EPA's Risk Management Program: How Does It Affect POTWs? EPA 550-F98-007. 1998

RMP Series. EPA's Risk Management Program: How Does It Affect Propane Retailers and Users? EPA 550-F98-008. 1998

Water Environment Federation and American Society of Civil Engineers. *Odor Control and Wastewater Treatment.* 1995

7. How Do I Comply With the Emergency Planning and Community Right-to-Know Act Regulations?

International Institute of Ammonia Refrigeration (IIAR). *IIAR Ammonia Refrigeration Education & Training Program: Basic Ammonia Refrigeration.* 1200 19th Street, NW, Suite 300, Washington, DC 20036-2412.

U.S. Environmental Protection Agency.

Report to President Clinton. Expansion of Community Right-to-Know Reporting to Include Chemical Use Data: Phase III of the Toxic Release Inventory.

Toxics Release Inventory: Reporting Requirements. Accessed via the Internet on October 28, 1997. http://www.epa.gov/opptintr/tri/rptreq.htm

Chemicals in Your Community: A Guide to the Emergency Planning and Community Right to Know Act. Chemical Emergency Preparedness and Prevention Office (CEPPO). EPA550-K-93-003. September 1988. Accessed via the Internet at http://www.epa.gov/swercepp/pubs/chem-com.html#P2fire.

Section 313 Emergency Planning and Community Right-to-Know Act: Guidance for Food Processors. Office of Pesticides and Toxic Substances. EPA 560/4-90-014. June 1990.

1992 Toxics Release Inventory Public Data Release. Office of Pollution Prevention and Toxics. 1992.

1994 Toxics Release Inventory Public Data Release. Office of Pollution Prevention and Toxics. 1994.

"Rankings of Three-Digit SIC Categories for the Food and Kindred Products Sector of 1992 TRI Releases." *Analysis from OECA's Targeting Work Group.* March 1995.

Emergency Planning and Community Right-To-Know Section 313: Guidance for Reporting Aqueous Ammonia. Office of Pollution Prevention and Toxics. EPA 745-R-95-003. July 1995.

EPCRA Section 313 Reporting Guidance for Food Processors, Office of Prevention, Pesticides and Toxic Substances. EPA 745-R-98-011. September 1998.

Walsh, James L. and Foley, Carol C. *An Update of SARA Title III: Toxic Release Inventory Reports for the Food Processing Industry - 1991.* 1991 Food Industry Environmental Conference. Georgia Tech Research Institute. Atlanta, Georgia. 1991.

8. How Do I Comply With the Hazardous Waste Regulations?

U.S. Environmental Protection Agency.

Land Disposal Restrictions Phase IV; Final Rule. Federal Register: May 12, 1997 (62 FR 91).

Doing Inventory Control Right For Underground Storage Tanks. Office of Solid Waste and Emergency Response. EPA 510-B-93-004. November 1993

Manual Tank Gauging For Small Underground Storage Tanks. Office of Solid Waste and Emergency Response. EPA 510-B-93-005. November 1993

Don't Wait Until 1998: Spill, Overfill, and Corrosion Protection For Underground Storage Tanks. Office of Solid Waste and Emergency Response. EPA 510-B-94-002. April 1994.

Environmental Regulations and Technology: Managing Used Motor Oil. Office of Research and Development. EPA/625/R-94/010. December 1994.

Understanding the Hazardous Waste Rule: A Handbook for Small Businesses - 1996 Update. Office of Solid Waste and Emergency Response. EPA 530-K-95-001. June 1996.

Closing Underground Storage Tanks. Office of Solid Waste and Emergency Response. EPA 510-F-96-004. August 1996.

Managing Used Oil: Advice for Small Businesses. Office of Solid Waste and Emergency Response. EPA 530-F-96-004. November 1996.

9. How Do I Comply With Spill or Chemical Release Requirements?

U.S. Environmental Protection Agency.

Chemicals in Your Community: A Guide to the Emergency Planning and Community Right to Know Act. Chemical Emergency Preparedness and Prevention Office (CEPPO). EPA550-K-93-003. September 1988. Accessed via the Internet at http://www.epa.gov/swercepp/pubs/chem-com.html#P2fire.

Understanding Oil Spills and Oil Spill Response. Office of Emergency and Remedial Response, Emergency Response Division. Publication 9200.5-105. EPA 540-K-93-003. October 1993

Risk Management Planning: Accidental Release Prevention -- Final Rule: Clean Air Act Section 112(r). Office of Solid Waste and Emergency Response. EPA 550-F-96-002. May 1996.

10. Other Major Environmental Statutes and Regulations: CERCLA, RCRA Subtitle D, FIFRA and TSCA

Kashmanian, Richard M., *Food Production and Environmental Stewardship: Examples of How Food Companies Work With Growers,* U.S. EPA, Office of Policy, Planning, and Evaluation. EPA 231-R-98-00. January 1998.

New York State College of Agriculture and Life Sciences at Cornell University. *Pesticide Applicator Training Manual Category 7: Industrial, Institutional, Structural, and Health Related Pest Control -- Food Processing.* March 1978.

Parmley, Mary Ann. *Modernizing Meat Inspection.* U.S. Department of Agriculture.

Ristaino, Jean B. *Agriculture, Methyl Bromide, and the Ozone Hole: Filling the Gaps.* AAAS/EPA Environmental Science and Engineering Fellow. Summer 1996.

U.S. Department of Agriculture. Food Safety and Inspection Service. *Almost Comprehensive List of Water-Related Issuances*. June 1992.

U.S. Department of Agriculture - Food Safety & Inspection Service. Protecting the Public from Foodborne Illness: The Food Safety and Inspection Service." *FSIS Backgrounder*. January 1995.

U.S. Department of Agriculture - Food Safety & Inspection Service. "A Farm-to-Table Food Strategy." *FSIS Key Facts*. January 1995.

U.S. Department of Agriculture. "USDA Unveils Sweeping New Food Safety Proposals." *NEWS*. Office of Communication. Release No. 0072.95. January 31, 1995.

U.S. Department of Agriculture - Food Safety & Inspection Service. "FSIS Pathogen Reduction/HACCP Proposal." *FSIS Backgrounder*. February 1995.

U.S. Environmental Protection Agency.

FARMFERT. Computer software providing guidance on proper handling and storage of pesticides, available at http://www.epa.gov/grtlakes/seahome/farmpest.html. 1991

Worker Protection Inspection Guidance, EPA 722-B-94-002, Office of Prevention, Pesticides, and Toxic Substances. January 1994.

Instructions for Reporting for the 1994 Partial Updating of the TSCA Chemical Inventory Data Base, Office of Prevention, Pesticides, and Toxic Substances, EPA 749-K-94-001. June 1994.

Part II: U.S. Department of Agriculture - Food Safety and Inspection Service. 9 CFR Part 308 et al. *Federal Register*. February 3, 1995.

"Raw Agricultural and Processed Commodities and Feedstuffs Derived from Field Crops." *Pesticide Assessment Guidelines Subdivision O Residue Chemistry Table II*. Office of Prevention, Pesticides and Toxic Substances. September 1995.

11. Pollution Prevention Techniques

General

Barlett, Kerry L., Lester, Richard R., and Pojasek, Robert B. *Prioritizing Pollution Prevention Opportunities With Activity-Based Costing*. Pollution Prevention Review. Cambridge: Cambridge Environmental Inc. October 1995.

Carawan, Roy et al. "Spinoff on Fruit and Vegetable Water and Wastewater management, " presented in *Industrial Water Conservation References of Food Processing*, California Department of Water Resources. 1989.

City of Los Angeles Department of Public Works. *Port of Los Angeles Pollution Prevention Project: Fish Processing/Canning Facility Assessment.* Prepared for City of Los Angeles Department of Public Works, Hazardous and Toxic Materials Office, and U.S. Environmental Protection Agency, Region IX. April 1994.

Crowley, Joseph C. and Sperber, Bob. "At KGF, Efficiency Meets 'Environmental Correctness'."

Delaware Department of Natural Resources and Environmental Control. *A Pollution Prevention Guide for Food Processors.* Pollution Prevention Program. Dover, DE., 19903. 1994.

Henry, Donald L. *Food Processing: Selecting a Site for Tomorrow.* 1996.

Jones, Harold R. *Pollution Control in the Dairy Industry.* Park Ridge, NJ: Noyes Data Corporation. 1974.

Northeast Waste Management Officials' Association (NEWMOA). *Pollution Prevention and Profitability: A Primer For Lenders.* 1996.

Purdue University. *Computer Integrated Food Manufacturing Center.* Accessed via the Internet at http://www.foodsci.purdue.edu/cifmc/publications/brochure.html

Robillard, Paul D. And Elliot, H.A. *Waste Management and Water Conservation in the Food Processing Industry. Proceedings of the Food Processing Waste Management and Water Conservation Conference.* November 1989.

Spitzer, Martin A., Pojasek, Robert, Robertaccoi, Francis L., and Nelson, Judith. *Accounting and Capital Budgeting for Pollution Prevention.* Pollution Prevention Staff. U.S. Environmental Protection Agency. January 13, 1993. Paper presented at the Engineering Foundation Conference, "Pollution Prevetion -- Making It Pay: Creating a Sustainable Corporation for Improving Environmental Quality," San Diego, CA. January 24-29, 1993.

U.S. Department of Agriculture. "Guidelines for the Safe Reuse of Treated Effluent Water for Meat and Poultry Processing." *Report of Water Guidelines Task Force.* June 1990.

U.S. Environmental Protection Agency - New England. *Financing Pollution Prevention Investments: A Guide for Small and Medium-Sized Businesses.* Undated.

U.S. Environmental Protection Agency and U.S. Agency for International Development. *Manual: Guidelines for Water Reuse.* EPA/625-R-93-004. September 1992.

U.S. Environmental Protection Agency.

Wise Guides: Energy Efficient, Cost Effective Actions for Food Processors. Climate Wise Program. Undated.

Overview of Environmental Control Measures and Problems in the Food Processing Industries. Office of Research and Development. EPA-600/2-79-009. January 1979.

Economic and Environmental Impact Case Studies: Food Processing. Climate Wise. Office of Policy, Planning, and Evaluation. Undated.

Pollution Prevention Case Studies Compendium. EPA/600/R-92/046. April 1992.

Environmental Research Brief: Waste Minimization Assessment for a Dairy. Office of Research and Development. EPA/600/S-92/005. June 1992.

Environmental Research Brief: Waste Minimization Assessment for a Manufacturer of Corn Syrup and Corn Starch. Office of Research and Development. EPA/600/S-94/016. September 1994.

Review of Pollution Prevention and Waste Minimization Techniques Used by the Meat and Poultry Processing Industry. September 1994.

Technology Transfer. Office of Research and Development. EPA/600/N-94-013. October 1994.

Case Study: A Better Tasting Pizza Sauce? NICE[3] Industry Uses an Energy-Saving Osmotic Filtration System. March-April 1995.

Food and Kindred Products Package- Poultry, Dairy, and Shrimp. Pollution Prevention Information Clearinghouse/ 1995

Food Processing: A Sector Study on Potential to Influence Environmental Sustainability of Food Production. April 1996.

Pollution Prevention News. Office of Pollution Prevention and Toxics. EPA 742-N-96-006. October-November 1996.

Excellence and Leadership in Environmental Protection. EPA 231-F-97-001. March 1998.

University of Wisconsin-Extension. *Small Business Waste Reduction Guide.* Solid and Hazardous Waste Education Center. September 1996.

Washington State Department of Ecology.

Pollution Prevention in Fruit and Vegetable Food Processing Industries. #94-56. March 1994.

Food Processor: A Success Story in Commercial Recycling and Waste Reduction. Solid Waste Services Program. May 1994.

Residuals Management

B., J. "Juice Company Recycles Residuals." *Biocycle.* November 1991.

Brandt, Robin C. and Martin, Kelli S. *The Food Processing Residual Management Manual.* Pennsylvania Department of Environmental Resources. 2500-BK-DER-1649. 1994.

G., J. "Recycling Industrial By-Products in Vermont." *Biocycle*. September 1991.

Robillard, Paul D. et al. *Recycling Food Processing Wastes in Pennsylvania: A Summary Report for the Ben Franklin Partnership Program*. Pennsylvania Food Industry Council. August 1992.

S., R. "Turning Residuals Into Bioproducts." *Biocycle*. November 1992.

U.S. Environmental Protection Agency. *Significance of Food Processing By-Products as Contributors to Animal Feeds: Phase I Food Processing Survey*. Prepared by the Office of Pesticides Programs, by the following: Rose, Walter W., Pedersen, Leo D., and Redsun, Harold, National Food Processors Association, Dublin, CA., and Butner R. Scott, Battelle Pacific Northwest Laboratories, Richland, WA. Contract Number 68-802-42643, EPA/HED #8. October 1989.

Composting

Biocycle.

"Blending Compost With Fertilizers." February 1993.

"Regional Composting of Waste Paper and Food." January 1995.

Brinton, Richard. "Low Cost Options For Fish Waste." *Biocycle*. March 1994.

Criner, George K., Kezis, Alan S., and O'Connor, John P. "Regional Compositing of Waste Paper and Food." *Biocycle*. January 1995.

Grobe, Karin. "Composter Links Up With Food Processor." *Biocycle*. July 1994.

Kashmanian, Richard M. "Poultry Industry Finds Added Value in Composting." *Biocycle*. January 1995.

Logsdon, Gene. "Turnaround in the Poultry Industry." *Biocycle*. February 1993.

Lowe, Eric D. and Buchmaster, Dennis R. "Dewatering Makes Big Difference in Compost Strategies." *Biocycle*. January 1995.

Ritchie, James D. "Fertilizing With Cheese Plant Sludge." *Biocycle*. February 1992.

Zorzi, G. et al. "Cocomposting Industrial Biomass." *Biocycle*. June 1992.

State Activities

Michigan Departments of Commerce and Natural Resources. *Waste Reduction Checklist*. Office of Waste Reduction Services. #8905. December 1989.

Minnesota Office of Waste Management.

Feeding Food By-Products to Livestock. Minnesota Technical Assistance Program. 4/91-77.

Source Reduction and Disposal Alternatives for Commercial Food Producers. Minnesota Technical Assistance Program. 4/91-76.

Waste Source Reduction Checklist. April 1991.

Minnesota TAP. *MNTAP Fact Sheet: Composting and Landspreading Commercial Food Wastes.* 1994. Accessed via the Internet on September 16, 1996.

New York State Department of Sanitation. *Blue Ridge Farms, Inc. - Waste Prevention Recommendations and Facility Assessment Report.* October 1996.

New York State Energy Office. *NYS Industrial Competitiveness: Spiral Belt Freezer.* February 1995.

North Carolina Agricultural Cooperative Extension Service. North Carolina Pollution Prevention Pays Program.

Bank or Drain: Cut Waste to Reduce Surcharges for Your Dairy Plant. CD-26. March 1996. http:\\www.bae.ncsu.edu/baeprograms/extension/publicat/wqwm/cd26.html

Bank or Drain: Dairy CEO's: Do You Have a $500 Million Opportunity? CD-29.

Bank or Drain: Liquid Assets for Your Dairy Plant. CD-21.

Bank or Drain: Liquid Assets for Your Poultry Plant. CD-20.

Bank or Drain: Pollution Prevention in Shrimp Processing. CD-25.

Bank or Drain: Poultry CEO's: You May Have a $60 Million Opportunity. CD-24.

Bank or Drain: Survey Shows That Poultry Processors Can Save Money by Conserving Water. CD-27.

Bank or Drain: Using COD to Measure Lost Product. CD-38. July 1991.

Bank or Drain: Using Food Processing By-Products for Animal Feed. CD-37. August 1991.

Bank or Drain: Water and Wastewater Management in a Dairy Processing Plant. CD-28.

North Carolina Department of Environment, Health, and Natural Resources.

Pollution Prevention Tips: Managing Food Preparation Wastes. May 1993.

Pollution Prevention Tips: A First Step to Waste Reduction: The Six-Step Waste Audit. October 1993.

Pollution Prevention Program: Case Studies A compilation of successful waste reduction projects implemented by North Carolina businessess and industries. " Case Studies SIC 2000: Food and Kindred Products." Pollution Prevention Program, Office of Waste Reduction. December 1994.

Ohio Environmental Protection Agency. Office of Pollution Prevention.

Pollution Prevention Assessment for Luigino's, Inc., Jackson, Ohio. January 1993.

Pollution Prevention Information Available From Ohio EPA. January 1995.

Oregon Department of Energy.

"Plate Cooler Keeps Bacteria Down, Profits Up." *Business Focus.* 330/55. 9-88/1000.

"Styton Canning Cuts Costs With Energy Efficiency." *Business Focus.* 330/26. 12-86/1000.

"Tax Credit Helps Carnation Cut Costs." *Business Focus.* 330/9. 9-86/500.

Innovative Technology

Environmental & Productivity Technology Innovation for the Food Manufacturing Industry. *Reduction of Food Processing Waste Stream Volume Through Membrane Filtration -- ET-2-A-(1).* May 1996.

Food Manufacturing Coalition. *Solutions to Environmental Quality & Productivity Problems Through Technological Innovation in the Food Manufacturing Industry.* Accessed via the Internet at http://ceres.esusda.gov:80/fmc/

Voluntary Programs

U.S. Environmental Protection Agency.

The Climate is Right for Action: Voluntary Programs to Prevent Atmospheric Pollution. EPA 430-K-94-004. August 1994.

Partnerships In Preventing Pollution: A Catalogue of the Agency's Partnership Programs. Office of the Administrator. EPA 100-B-96-001. 1996.

www.ingramcontent.com/pod-product-compliance
Lightning Source LLC
Chambersburg PA
CBHW080633180526
45168CB00008B/3159